Packaging Design

Howard Milton

The Design Council

Published in the United Kingdom in 1991 by

The Design Council

28 Haymarket

London SW1Y 4SU

Printed and bound in the United Kingdom by

Bourne Press Ltd, Bournemouth

Reprinted 1996

Designed by Nicole Griffin

British Library Cataloguing in Publication Data

Milton, Howard
Packaging Design
   1. Consumer goods. Packages. Design
   I. Title II. Design Council III. Series
   688.8

ISBN 0 85072 280 2

Cover: Jif Lemon, the epitome of product and packaging in harmony.

Contents

Acknowledgements

The author and publishers gratefully acknowledge permission to reproduce the following illustrations: Robert Opie Collection for 1001 Carpet Cleaner (page 4) and Typhoo Tea (page 56 lower); Charlie Stebbings for Jif Lemon (cover) and Natrel (page 79;); Martin Barraud for 1001 Carpet Cleaner (page 5), Dulux Paint (page 49) and Bassett's Allsorts (page 83); Francis Lumley for Cluster Bar (page 40); Paul Williams for Pedigree Chum (page 53); Mark Finkensteadt for Typhoo Tea (page 56 upper) and Toilet Duck (page 59); Trevor Key for Radion (page 87); Paul Dunn for Heinz Tomato Ketchup (page 95).

The author wishes to thank Gillette and Bassett's for their detailed contribution to the case studies; Greg Sheridan and Sophie Caruth for their research skills; Jay Smith for her focus; and Lucille Barha for her patience at the keyboard.

The author

Howard Milton was introduced to drawing by Ruskin Spear and to design by Tom Eckersley. After leaving the London College of Printing in 1974 he was hired by Michael Peters and he began to pioneer his principles of the pack personality. Following a year in advertising in New York, Howard returned to London and co-founded the design consultancy Smith & Milton which, since the early 1980s, has been at the forefront of creative packaging.

Preface

This book is written as background reference for anybody who is responsible for considering and implementing packaging design decisions – brand managers and designers alike. It focuses on the function of packaging design in establishing and maintaining product and brand identity.

Packaging design is different from virtually every other design discipline because its primary objective is simply recognition. Successful packaging design, typified by the major manufacturing brands, calls first and foremost for consistency. Packaging becomes a symbol of identity that is constant, familiar and instantly recognizable. Throughout this book I will return to the central theme of continuity.

Essentially there is too much packaging. The speed with which new packaging design is developed and old design is dumped compounds directly on the amount of natural resources exploited and the physical packaging wasted. If a change of packaging design really is required, then how can the designer help to minimize any harmful effects on the natural environment? It is easy to recommend recyclable board, biodegradable plastics, or refillable packs. But any decision to switch to new materials rests with the client company and its managers and it is to them that I make an appeal for more responsible attitude towards the environment.

I refer readers who are about to work with designers for the first time to Chapters 8 and 9. Working with creative people should be a stimulating experience, so why do so many clients complain of frustration and dissatisfaction? The relationship between manufacturer and consultancy should be one of depth and commitment, founded on mutual trust and rapport and not on slick artwork. Chapter 9 sets out to prove that briefing, the basis of good design, is a two-way process between client and designer.

Finally, I make no apologies for borrowing my approach from another personal passion – football. The lesson I learned from my football heroes was to be competent in the basic skills; first learn control and then play the simple ball. These were essentials to be tackled before any flamboyant moves could be attempted. I feel that the leading examples of brand awareness suggest a very similar approach: master the basics, play the simple game, and keep your discipline even in adversity.

Howard Milton
January 1991

1 Making the right impression

In the average brand manager's day, very little time is actively devoted to design management. Packaging design is often the neglected ingredient in the marketing mix, even though the brand's recognition is fundamental to its success and critical to its promotion. By briefly highlighting snapshots of brand history, I am seeking here to convey the potential of a brand's visual and emotive origins, and to focus its future appearance by realizing its potential.

▪ Product recognition

The idea of cleaning carpets is unlikely to be uppermost in most people's minds. Even so, after many years, I can vividly recall the TV advertising image of a hand holding a bottle of carpet cleaner and hear the slogan '1001 cleans a BIG BIG carpet for less than half-a-crown!'. At the age of ten I had little interest in advertising and even less in carpet cleaning. So why did the image have such impact? Certainly I might have seen my mother using 1001 and would have seen the label for more than a few seconds on TV. But would I remember it so clearly if its promise had not been phrased so forcefully? The answer lies in the dual strength of a simple proposition, executed to precision, and an equally understandable graphic – 1001 – which created an indelible impression.

Through simple, bold graphics 1001 asserted an enduring identity

Today, despite the fact that it has not made a reappearance in any TV commercial or national advertising campaign since 1967, 1001 is still going strong. It sells almost six million units of liquid a year and is the leading carpet cleaning brand with a market share of 45 per cent, nearly three times greater than at the height of its TV exposure in the late 1960s. The strength of 1001's brand image effectively killed off any rivals, allowing the manufacturer, Cussons (perhaps more famous for its soap), to introduce other products (a foam, a stain remover and a version suitable for machine use) as natural partners to the original 1001 product.

Since the mid-1960s, the 1001 brand identity has undergone considerable cosmetic change. A PVC container with built-in hand grip has replaced the original glass bottle, and the utilitarian, albeit very British, red, white and blue label has been banished in favour of a more illustrative, less brash counterpart. The product is priced at substantially more than half-a-crown, but the proposition of bigness and efficacy remains wholly intact.

New materials and a modulated graphic style have assisted 1001's market lead

1001 is a great survivor, a product which, through a strong but simple offer, staked its claim as a brand unique in personality. It is also a rare product, one whose success has opened up other markets and which, without serious competition, has effectively been positioned as the generic brand, the first choice for tackling specific cleaning tasks.

The majority of products on the supermarket shelf have, however, a very different background. Locked between the loyalty of the consumer and the demands of the retailer, they are fighting for visibility, shelf space and market share.

Brand marketing is a volatile business in today's highly

competitive retail environment. With the continuing emphasis on 'creative' advertising and the 1980s' crop of 'creative' packaging designers, the once gentle evolution of a product's image has accelerated at a frenetic pace. There are more products available, so consumers have more choice and yet, simultaneously, are more discerning. The packaging of goods has to work harder than ever if the product is to be noticed and survive. While most brand managers were formerly content to count the tonnes and keep a wary eye on the nearest rival, competition has intensified to the point where no one can afford to be complacent.

The rise and success of retailers, for so long subservient to the leading manufacturers' brands, have shifted the power balance and created new threats with intensified pressure. Throughout the 1980s, the decision by manufacturers to repackage a product was forced by the breakneck speed at which own-brands were appearing, and because the leading multiples had the ability to conduct on-shelf research, allocate shelf space as they liked and at the exclusion of whom they chose.

A product's viability is no longer assessed over a period of years. It now has only a matter of months, sometimes weeks, in which to prove itself. For example, new products introduced into the Marks & Spencer range of groceries are given two weeks in which to sell. Unsuccessful ones are dropped. Therefore, as a communications medium, packaging is the stage upon which the product must perform. A bad performance will not invite an encore.

For every product as successful as 1001 there are a hundred more casualties, either because the product was not wanted or needed, or because it failed to communicate the right message to the right audience.

Timing the launch of a new product is critical, and depends on meticulous research into the target market. Few manufacturers, for instance, would have extolled the environmental virtues of a new, mass market product in the 1950s or 1960s, an aspect of manufacture that, currently, none can ignore.

Despite shifts in attitude, what makes a person select one product in preference to another?

▌Getting the message across

Early packaging concentrated on the product promise for its effect. The surest way of emphasizing a product's difference was to spell it out clearly. This led to the birth of the mark of difference, the symbol of recognition known as 'the brand'. But once a brand's popularity was established, attention switched to broadening its market and the effort in promoting the brand took precedence.

As the advertising medium provided a larger canvas, it created the opportunity to use persuasive images as well as words to add a visual and emotional dimension to many products. But many advertising campaigns developed to the exclusion of a product's on-shelf appearance, frequently glorifying the virtues of taste, form or function in obscure or esoteric ways. In the early part of this century, the consumer was bombarded with product benefits which were often far-fetched:

- Beecham's pills ensure good health   (1911)
- Bovril for health, strength and beauty   (1930)
- For your throat's sake, smoke Craven 'A'   (1936)
- Fry's Cocoa – Rich in nerve food   (1940)

Legislation has since helped to reassure the consumer that the content of any advertisement is 'legal, decent and truthful'. Within these necessary limits, packaging has to perform very much harder to attract the consumer who is confronted with a vast array of products, all advertising the same guarantee of performance, quality or satisfaction.

Creating a pack design that imparts a total product story has to be attempted in a single statement. Packaging design is at its most effective as a selling vehicle if that message is endorsed consistently through its advertising, a highly visible medium. The evolution of the advertising campaign for a brand should be inseparable from the primary proposition, the product. However, in some cases the use of dominant and influential media – TV and cinema, for example – has, through the often esoteric nature of today's advertisements, obscured the real advantage of many products. Consumers might be entertained by a TV or film commercial but they may remain unable to identify the product or its qualities; 'I know the ad but I can't remember what it was advertising'.

If there is a great difference between a brand's advertising image and its physical on-shelf appearance this can create discontinuity in the mind of the consumer. People look for some visible link between what they have seen (and believed) and what they are buying. A 40-year-old brand may have metamorphosed many times, with each design change playing down some characteristics, creating new ones and enhancing others, but there must be some element of continuity to reassure the purchaser. For instance, the following copy line examples are now so familiar that consumers can automatically attach them to the product name and some people, no doubt, could also give a fair description of the pack graphics:

- Makes exceedingly good cakes
- Probably the best lager in the world

- Helps you work, rest and play
- Top breeders recommend it
- High in essential polyunsaturates

Not surprisingly, all these products are strong market leaders which have stuck to winning strategies.

Advertising which plays by simple rules remains the most memorable. The same is true of brand design. The aim is longevity. Once a strategy is discovered which endorses a product's performance or emphasizes its perceived advantage, it should be adhered to. Like the memorable copy lines imprinted on the mind of the consumer, impression and brand proposition should be inseparable. Whilst the method of promoting a product is changeable according to style and fashion, the product itself, and consequently its packaging must, by and large, act as the constant factor. This implies that the pack is easy to recognize, establishes an emotive link with the advertised values and, above all, is a permanent visible communication. It is almost a hackneyed notion that the packaging is permanent advertising, but the fact remains. Whether emphatically pleading for attention in a store or reassuring by its presence at home, the pack image must work continuously.

Golden Shred: a striking and memorable identity little changed for 50 years

Surely it is no coincidence that the most memorable of packaging images – Brasso, Golden Shred and Kit Kat, for instance – are the least changed for over fifty years. At what cost to customer loyalty have

Campbell's soup, Sunsilk shampoo and Maxwell House coffee paid by their constant redressing in an effort to appear more delicious, fashionable or visible? Whilst each of these, and countless others, originally possessed strong and individual identities, it may be argued that a mismanagement of the visual elements in conjunction with a fluctuating brand message has succeeded only in dissipating any lasting impression.

Reaching for a redesign every two or three years is not the strategy to secure brand awareness. Likewise the panic riposte to a new market entry encourages further dissolution of a brand's integrity, and is the recipe for ultimate anonymity. Packaging design can take the rap for a decline in sales only when guilty of being created in isolation. Without support, few products could grow to a size that would give them a real chance to dominate a sector of the market, but, as 1001 exemplifies, the establishment of a clear product benefit can endure beyond the time-span of an advertising campaign. Whilst label design has altered the tone of voice of the current pack, the inherent strength of the brand remains the unequivocal stance of the 1001 symbol, which owes little of its appeal to beauty. Its strength remains its simplicity as a visual and aural mnemonic.

2 Brands and manufacturers

Recognizable branding is the means by which consumers can identify a product or service which offers specific benefits. A brand distinguishes itself from a product because it possesses an identity which signals to the consumer four elements: purpose and performance, quality and value.

Brand names and packaging identities are now part of everyday life. Our ease in differentiating between, say, a Ford and a Rolls-Royce car, is not built solely on the physical difference, but on the images they convey. For most people it matters little that Ford is American, or that Rolls-Royce also manufactures jet engines; these are not the images that are primarily recognized or projected.

Through this recognition, brands become representative of product areas and establish product sector language. So Kellogg's, famous for breakfast cereals, could conceivably also bake bread. To the consumer, however, it would be totally implausible for Kellogg's to manufacture washing powder.

Brands are built on years of consumer understanding, through advertising, design and other forms of promotion. Brands are a manufacturer's most valuable asset. Research carried out in 1989 by Payne Stracey Partners for the international branding consultancy Interbrand showed that 88 per cent of the UK's top finance directors believed that brands have a value which should be measured and known; and 83 per cent further believed that these valuations should become part of a company's financial statements. Identifying and assessing a brand's worth is dependent on the number of consumers in whose minds it is embedded, but it is a highly subjective exercise and one which the City has had great difficulty in coming to terms with. It is, after all, difficult to appraise the value of attitude as if it were a commodity. The City's failure to appreciate fully the value of brands was highlighted in 1988 by the lack of support given to Rowntree prior to its takeover by the Swiss organization Nestlé, which paid £2.2 billion for famous names such as Kit Kat, Quality Street, Smarties, Polo and After Eight.

▮ Company versus brand

If a brand name is well known and a new product is being primed for introduction to the market, should a manufacturer launch it on the back of a recognized brand or brand it uniquely? Is it the company which makes the brand or the brand which makes the company? Whereas the ever-popular Kit Kat, Smarties and Aero are not immediately identifiable as Rowntree products, there is no mistaking Cadbury's Caramel, Wispa or Creme Egg. Rowntree and Cadbury's are two equally successful organizations operating within the highly price-sensitive confectionery industry, yet they employ very different marketing strategies. The individual brand name approach of Rowntree focuses solely on product differentiation and allows a broad product range. An amalgamation of Rowntree's sugar confectionery skills and Mackintosh's toffee expertise means the joint company now offers confectionery for every taste, from extra strong mints through to chocolate wafers. Cadbury's, on the other hand, remains largely focused on its chocolate origins, and it is the company's long-standing expertise within this area that forms the basis of each new and existing product. Similar situations exist within other market sectors.

For a new company launching its first product into a crowded marketplace the owner's name often takes precedence over the brand name. Whether offering a better version of an existing product or launching the rare, original claim, a manufacturer's name has traditionally signalled the first point of difference. There are, of course, notable exceptions. Steve Jobs, frustrated at being unable to decide upon a name for his newly founded computer company, vowed to call it anything. In the end, the apple he was eating at the time provided the inspiration.

Branding and design decisions present themselves at an early stage in a new product's life. Should a company follow a corporate stance or opt for flexibility through individual product identities? This is a fundamental dilemma encompassing four basic choices which can be identified as corporate style, house style, range identity or individual branding. Although each is distinct, many companies create brands which extend into two or three of the categories.

Corporate style

The brand name is used as a collective where individual products are less well-known than the operating function of the company. This gives organizations such as Ford, Apple, Adidas and Clark's the flexibility to launch a new product at any time (within their recognized product sectors), in the knowledge that it will be associated with the quality of the corporate brand. Since the launch of the Apple computer, for example, the company has introduced computer-related products from add-on hardware to software, all bearing the company's stamp, which to the consumer represents a guarantee of product performance.

House style

House style is often adopted by companies whose products cross many market sectors, and therefore require a unifying link back to the central proposition. The St Michael brand, for example, is harnessed by Marks & Spencer to endorse the quality and no-quibble guarantee of the entire range of products that are unique to its stores.

The benefits of this type of branding tend to be self-reinforcing, and are regularly used in advertising. So a favourable response in one product area, such as Heinz Baked Beans, may encourage consumer confidence in other Heinz products – babyfood, for example.

Range identity

Range identity encompasses brands which span a number of products, linked by a core benefit. Mr Kipling's 'exceedingly good cakes' epitomizes a range which is indefinitely extendable within the bakery sector, and is a strength created independently of any knowledge of the manufacturer's real identity: Rank Hovis McDougall. Likewise, Dulux's own personable image and its stronghold in the home decorating and colour paint market divorce it from its comparatively faceless parent, ICI.

Individual branding

Originally defined by a single product within a single market, individual branding includes flavour, size and format options. Some of the best-known brands, such as Coca-Cola, Mars, Elastoplast, Flora, Persil and Bisto, fall within this category. Coca-Cola, Mars and Perrier are anomalous, however, for each is a company in its own right and carries the weight of the corporate brand. Mars also markets products using a range identity such as Pedigree petfoods, which in turn runs individual brands, such as Whiskas cat food. The strength of each of these brands, whether corporate or individual, lies in the ease with which the consumer can now confidently identify them and judge their quality.

Assumptions made by the consumer based on appeal, quality and performance derive from the relationship and understanding that the consumer has established with a name and an impression. Achieving that confidence is a manufacturer's key to success. The supposition that a competitor's bottle of tomato ketchup, for instance, could match the quality and taste expectation of Heinz is hard to imagine. But the fundamental rules of marketing necessitate choice before preference.

Whilst the power of the disciplined pack has added to the leading position of such brands, the opportunity for choice has created a new identity category.

∎ Mainstream own-label

Own-label is a familiar concept in supermarkets and department stores. It used to be the poor relation of branded goods. The consumer was under no illusion that an own-label product was likely to be cheaper than and, by implication, inferior to the market leader, despite the tendency to mimic the brand's appearance.

During the 1970s, the image of own-label products was so poor that they were virtually withdrawn from the marketplace. Where previously they had aped the leaders' livery and fashioned their image as close to the established sector symbols as they dared, own-label products began to shy away from design. What became known as 'generic' packaging, austere white or yellow packs with no evident graphic design, relying instead on uninspiring product descriptions was launched: for example, '20 economy dustbin liners', or the equally deadpan 'Baked beans in tomato sauce'. Visually there was no contest between the glamorous brand leader and the bland own-label alternative. But in the lean years of the 1970s, price rather than aesthetics was the shopper's chief concern. In the 1980s the return to economic prosperity and optimism led to a turnaround in attitude, instigated by the leading grocery multiples who not only freed themselves from paying lip service to branded goods, but reinvented their own brands in order to compete directly with manufacturers.

Retailers have not only repositioned their products, they have also invested in and launched new ones. Now own-label has become own-brand and with it has come the necessity for more shelf space. Take Sainsbury's, for example, whose brand image has long represented the benchmark against which most other supermarkets measure their own success. Even with the presence of strongly supported branded products such as coffee, Sainsbury's continues to dominate whole product sectors and remains the sales leader within its owns stores.

In the USA competition for shelf space and product visibility has become so intense that some retailers charge for shelf space as if it were real estate. As in the UK, manufacturers are obliged to make substantial investment in image-building and brand-enhancement programmes in order to ensure that their products remain uppermost in the consumers' minds, and consequently enjoy a rapid turnover at point of sale.

∎ Life cycles

Despite the pressure created by branded and own-brand competition, there is no doubt that it is product performance and not a new pack design which is the key to extended product life. Regardless of how visually sophisticated, aware and discriminating we believe ourselves to be, the fact remains that we are still searching for the reassurance that product benefit is derived from the product itself.

Areas deemed 'lifestyle' (such as boxed tissues and bath foam) are not exempt from the same discipline. If such a product aims to achieve longevity, it must establish a firm 'image' base and seek to enforce that value. Cussons Imperial Leather soap is a clear example of such a mainstream brand. Its livery owes little to today's high street trends, but its unrivalled market position is indicative of its success. Consequently the continuous quest for product superiority and difference drives manufacturer and retailer alike. The counterbalance to the proven brands and the retailer alternative is the ever-anticipated arrival of a new product with a new perspective. The genuinely new product (of which there are few) can virtually write its own visual and emotional language. Its only point of reference is the products it aims to replace, and its difference from those should be marked.

The chances of a newly launched product supporting itself long enough to make an acceptable profitable return are currently very slim. In the 1980s, in the USA, the estimated failure rate of new products launched in the grocery sector was a disheartening 90 per cent (Booz, Allen and Hamilton 1982). While this failure rate may seem reason enough to support a new product by association with existing brand successes, the converse argument for individual branding is that it provides a mask for any manufacturer should the new product (as is statistically likely) 'bomb'. Such pessimistic odds suggest that the consumer is either very loyal or very blinkered and that a manufacturer should think twice before investing in a new idea. But in reality new products or product developments are introduced constantly, often merely to allow products to maintain some presence.

The initial idea for and evolution of a major brand follow stages which can be tracked. From its inception, a product will follow a pattern of growth until it reaches maturity. It will reach saturation point and eventually it will go into decline. The analogy of the life of a brand with the human life cycle is closest at the beginning and end. Initial enthusiasm for a project will invariably assure that it receives wholehearted backing for its launch and its drive towards rapid growth. In time, a decline in interest might lead to a withdrawal of support which can only quicken the end of a product's life.

In the middle years of its life a product is never allowed to stand still. Increasing competitive activity, the introduction of new technologies and rapidly changing consumer tastes will all accelerate modification and change. The design may be fine-tuned to maximize a product's presence in the market. If targets are not met, an attempt to rekindle interest and inject life into an ailing product may be made by the manufacturer in the form of a redesign strategy. This is a common reaction often brought about by the challenge of a retailer's own-brands, and signifies the inevitable approach of old age.

Could it be that it is the product itself that is ageing, not the brand idea? Or has a fundamental attempt been made to shift the packaging appearance, which has left it devoid of those original signals that created memorability and reassured the consumer of the product's effectiveness? Given that fashion cycles are getting shorter and that design changes are usually an attempt to echo these, is the money not better spent banking on a product's past success or does this imply inertia? Perhaps it is worth thinking about some brands old enough to have been victims of such a life cycle, but who have come through, and are back for more.

The factors affecting a product's life and the reasons for effecting a packaging change are innumerable. It may be a defensive move in response to a rival product or a positive action prompted by the development of a new physical advantage. There are however, more dubious reasons for change, ranging from 'I'm new and I've got to make my mark while I'm here' through to the autonomous junior brand manager, who has packaging design within his or her power, but not the experience to foretell the consequences of his or her action.

No advertising campaign is ever treated so lightly nor its strategy left susceptible to short-term thinking. Packaging design, whether representing a leading brand or a new market entry, is not just a support for advertising, it is advertising. The more that consumers are made constantly aware of a product's existence the more likely they are to buy it. And the more identifiable the 'brand', the easier it is for the process of recognition and association to take place.

▍Brands that live on

The resourceful manager is adept at watching out on all fronts, and being aware not only of the product life cycle and the erosion of key markets, but also of the 'brand cycle' and the opportunities which exist within it.

It is possible to extend the life of a brand by transferring its core benefits to new products. Some of the UK's leading brands have maintained their position for over 50 years because they have been built on and are supported by a single brand strength. A survey conducted by the market research company Neilsen (1989) for *Checkout*, revealed that, on average, the UK's ten most popular brands are 42 years old. Unilever's 80-year-old Persil washing powder is Britain's top-selling grocery product with annual sales of £192 million, followed by 51-year-old Nescafé coffee at £188 million, and 30-year-old Whiskas cat food at £180 million. As with 1001 cleaner, which has been adapted to satisfy the needs of new markets (vinyl carpet owners, for example), brand image becomes synonymous with individual product areas. Hoover is to vacuum cleaners what Walkman has become to personal stereos.

The chart (below) lists some of today's most famous brands and, since the 1930s, the most successful. Whilst Heinz has been able to introduce new and interesting soup varieties throughout its 80-year existence it has maintained recognition through a red label and its famous keystone device that confirms familiarity and reassurance, with impact similar to the yellow Kodak film box. None of these brand leaders has stood still.

Brand	Position 1935	Position 1988
Bird's Custard	1	1
Heinz Soup	1	1
Stork Margarine	1	1
Kellogg's Cornflakes	1	1
McVitie's Digestives	1	1
Cadbury's Dairy Milk	1	1
Schweppes Mixers	1	1
Rowntree Fruit Pastilles	1	1
Kodak Film	1	1
Gillette Razors	1	1
Brooke Bond Tea	1	1

3 Brand awareness

For every new consumer a product attracts, it is just as likely to lose another consumer to a competing product. Shopping is as much a subjective as an objective activity. Of course, the majority of consumers take time and give some thought before forming definitive opinions about whether a purchase is worthwhile. It is at this point that packaging design can be most persuasive. Whilst advertising may alert a large number of potential consumers to a product's existence, it is only at the point of purchase that the promotional story and the product's image come together.

Few marketing managers will risk the success of their new product by allowing customers to make their first-time purchase decision on the strength of the packaging; that is, without the back-up of advertising or a special offer. Advertising generates product awareness, promotes or simply suggests the product's positive advantages, and creates the correct ambience for its chosen audience. A good example is Timotei shampoo, a product for everyday hairwashers, whose advertising evokes an impression of naturalness, through the use of images of a tranquil, Nordic nature-trail, and freshness and purity, through the central character of a blonde-haired sylph. Timotei became a clear brand leader in an extremely fickle market sector.

With an existing product, however, advertising's role is more specific. It has to address the consumer's existing preconceptions and work on the product's positive attributes whilst taking account of factors that may have created negative perceptions.

A change in packaging design is not going to produce an instant change in consumer feeling. Dressing for 'activity' or 'modernity' is merely cosmetic. Once an apprehension has been introduced over performance or quality, it requires fundamental restructuring to restore confidence in a product or company.

The discovery in 1990 of the potentially harmful benzine in the bottling technique of Perrier mineral water presented a major problem for the brand. Not only did the scare hit sales and prompt once-loyal customers to seek an alternative, but the revised filling method failed to satisfy Sainsbury's, one of the largest UK outlets for the product. Sainsbury's refused to stock Perrier whilst the label bore the claim 'naturally carbonated'. Such an embargo on any product does little for consumer confidence, not to mention the heavy promotional costs for its reappearance. However, the heritage of the Perrier brand in the UK is as much built upon its advertising style as is the product's fashionability. The possibilities to exploit the phonetic potential of the French word 'eau' have created some memorable and humorous puns. The temptation to switch to some 'worthier' product advertising during the crisis must have arisen, but the decision to stick to the familiar formula has lessened the impact on doubtful consumers, and will probably be a major contribution to a return to normal sales.

It is the role of advertising to assimilate impressions and project the product's positive qualities whilst eradicating any negative qualities. For a television commercial to make an impact, the brand image itself has to work fast, hard and make its impression first time round.

Inadequate coverage of product identity is wasted airtime and therefore money. The retention of such imagery is paramount in the building of a brand personality.

▌Brand damage

Coca-Cola is a lifestyle, symbolized as truth, freedom and the American way. Formulated in 1886, Coca-Cola became an international phenomenon during the Second World War when American GIs were sent overseas, taking with them everything American, including Mom's apple pie, chewing gum and Coke-was-it. The drink became synonymous with youth and affluence. For good measure the more multilingually acceptable name 'Coke' was adopted, a nickname that gave the brand an even friendlier profile and simultaneously protected it from its growing cola rivals. Subsequently the red roundel used in advertising and at the point of sale became so well known that its presence superseded the famous 'waisted' glass bottle. This facilitated the packaging transition from conventional glass to cheaper, more easily transportable tinplate, which in turn has evolved into environmentally conscious, recyclable aluminium. But crucially, in terms of its packaging evolution, it remained true to the distinctive scripted Coca-Cola logo and its attention-seeking red background. All the while, advertising majored on fresh-faced, happy youths growing up enjoying the drink. With emotions running so high the sensitivity required to navigate such a brand took on corporate proportions. Any decision was a corporate not just a marketing decision.

The launch of New Coke in 1985 is possibly the most publicized marketing disaster of recent years. The arch rival, Pepsi, through a massive, single-minded campaign, had confidently established that consumers preferred the Pepsi taste. This prompted Coca-Cola to reformulate its own product, one whose strength lay in a 99-year-old secret recipe. In response to consumer indications following a three-year tasting project, Coca-Cola was convinced that it had developed a new Coke with a new taste capable of outdoing Pepsi. The company was wrong. The notion that the 'original' Coke recipe had been changed provoked what can only be described as a national outcry and corporate panic, which resulted in the original recipe being relaunched as Classic Coke only three weeks after the New Coke launch initiative. In the ensuing turmoil, Pepsi continued to outsell both Classic and New Coke for over six months.

By not evaluating the power of the Coca-Cola ideal and everything it had signified for almost a century, the company was guilty of a fundamental error. It had overlooked the fact that the brand had been outmanoeuvred not because it suffered from an inferior taste, but because it had fallen victim to relentless marketing pressure from Pepsi. In terms of its cultural and emotional associations, Coca-Cola is a far stronger brand than Pepsi. Concentrated support of the old brand principles would probably have been sufficient to face up to the threat from Pepsi.

In America many people have still not recovered from the idea that the recipe was changed. Even in taste tests where the Classic Coke formula is identified, consumers still say that it does not taste as good as it used to.

An alternative strategy may have been for Coca-Cola to have launched an equally well-supported new brand which conceivably could have taken the pressure off the primary brand. However, in this instance it would not have been necessary. By concentrating on the existing brand's strength and by taking a positive decision not to change, Coca-Cola should have chanelled its resources into promoting the product's emotive power, not into fighting an opponent-manipulated weakness.

Building on the strength of existing brands is always an attractive option. Attempting to emulate the success of a first product is a common pitfall for many smaller companies. Since the success of a new product launch is largely reliant on its innovation, a second product following the same idea is likely to flounder. In other words, if a totally unique formula has been achieved with the first launch, a secondary product deviating only slightly from the first and not building upon the distinctiveness of the former becomes a largely redundant addition.

Missed opportunities occur when a company fails to produce a good second product whilst promoting the quality of the first. For big manufacturers, there is an opportunity to enter into markets where their core brand credibility could otherwise be questioned. The family tree of Mars, for example, is so little known to most consumers that they are unaware that the sweet bar they have just eaten originates from the same company that manufactured the dog's dinner.

The relationship between consumers and brands is highly charged with emotion. A positive experience with one product can prompt a consumer's confidence and trial in another. However, as in the case of Coke, if consumers' affections or loyalties are betrayed (if their perception is altered), the resulting damage can be both great and long lasting.

It is difficult to predict what emotional signals a consumer will pick up from a brand, since so much is preconditioned by the product's advertising support. However, once a product has been purchased its performance must match its advertising propositions. If the product itself disappoints, the brand has to perform what can only be described as an on-shelf miracle the next time around if it is to attract the same consumer again.

4 Inventing the proposition

Before any consideration is given to the graphic interpretation of a package, the first step must be to establish the desired response from the target audience. Whilst Coca-Cola, built over generations of image endorsement, operates *en masse*, the positioning of new products and the reaffirming of existing ones relies upon a clear understanding of the market's taste and preference. Talking to the audience created success for Coca-Cola; asking the wrong questions led to a near disaster.

Market divisions were once glibly classified as A, B, C1, C2, D and E, according to income and class. Since the mid-1970s this method of classification has become virtually redundant. Today there is a more focused method of defining or identifying the consumer, which groups people by their behavioural preferences rather than by their financial means. The specifics for any group can be assembled to match product or service. This pairing owes more to accepted taste values and allows for popularity across age bands as well as wage bands. Using this method of market segmentation, consumers can be identified by the shops they choose, the cars they drive, the newspapers they read and the films they prefer. The market 'identikit' is invaluable for products which are subject to fashion cycles because such products (household paint, for example) are likely to attract media coverage, which in turn encourages trial purchase. Peer-group influence is also likely to prompt purchases in this particular market. However, it is a common error to segment the market to the point that the target audience is too small to justify the expense of launching a product aimed exclusively at that audience.

∎ Niche marketing

Recognizing the opportunity to market to a very specific group of consumers is invariably undertaken by new or small businesses. The investment in 'difference', the knowledge that must be acquired to satisfy the exact requirements of a specific audience, invariably dictates high cost, owing to short-run manufacture and high product specification. These are factors that produce premium-priced products, but require a substantial return to ensure the company's survival.

Premium-priced products may produce high profit margins, but they also require skilful management if they are to penetrate successfully beyond the chosen target market. Provided that the research has been thorough and the market sector has been shrewdly judged, the likelihood of success for good niche products is high because they can command high consumer loyalty and repeat purchase.

Such markets grow and can attract larger manufacturers who are seeking further opportunities to develop a mass market. The opinion-forming consumers who established the niche possibility may stay loyal to their own product, but their lead can often influence the future tastes of the average consumer. W Jordan (Cereals) Ltd, a milling firm that has been in business for over a century, did just this. It developed a cereal product to appeal to health-conscious customers. By combining its basic produce of seeds, nuts, grain and fruit into a single confectionery bar format, it created a unique, 'healthy' snack bar which was low in sugar and free from artificial additives. Packaged in a 'natural' white sleeve, endorsed by semi-handmade stencil typography, the identity betrayed none of its humble origins.

Whilst satisfying its health-food conscious clientele, the Jordan bar also attracted a bigger market. As the desire for healthier eating increased, so too did the interest in snack bars from other companies such as Lyons-Tetley, Quaker and Mars. By the mid-1980s, these manufacturers had launched respectively Cluster, Harvest Crunch and Tracker, each with its own mass market-orientated packaging style. All three were more visible on shelf, and led with a brand name. The no-compromise, specialist stance of Jordan ensured its position within the bigger market but not without the inevitable change of packaging to match the more confident and noticeable designs of its imitators.

Jordan has taken more than a hundred years to change its market direction. It is difficult to guess what reaction the bar would have received if launched 20 or 30 years earlier. It is probable that the low awareness of healthy eating would have meant the product would have been evaluated solely on taste and would have lost out to the established confectionery of the day.

The development of companies like Jordan should be a lesson to the majority of marketeers who adhere to the market-share mentality, which is constrained to established markets where the aim is to capture and maintain a slice, no matter the price. The niche marketeer, on the other hand, creates new and potentially bigger markets. However, it is not simply 'opportunity' that governs successful new product launches. Unless niche marketeers thoroughly research the feasibility of the proposition they may reveal, through continually eliminating potentially unsuitable targets, that a market never really existed at all. The larger, mature manufacturers who seek to develop a new market will always back their judgement with thorough research. The guidelines for any new development can be mapped out successfully only if an in-depth understanding and assessment of the target consumers' fears and insecurities, as well as their desires, can be achieved. Very often a negative consumer attitude creates a positive product (Davidson 1987).

▌Dulux: soft shades of white

Wall paint is a good example of a fashion product. It is relatively cheap, easy to use, and can dramatically change the mood or feel of a home. But people are terrified of it. Colour is one of the most difficult areas of 'lifestyle' to exploit. Despite the ability to choose ties, socks, curtains and carpets, most people lose confidence when faced with decisions about what colour to paint a bare wall.

Towards the end of the 1970s, Dulux, the market leader in white and white-and-coloured paints, was under pressure from Crown, its biggest competitor, and a rising number of own-label paint ranges. An earlier strategy had seen Dulux launch an 'improved' white paint under the name of 'Pure Brilliant White'. This resulted in increased sales but it was soon challenged by the launch of similarly named rival products.

Dulux conducted extensive research into the white paint sector and discovered the 'brilliant' opportunity. Whilst it was already selling white paint at premium cost to reflect quality, this did not take account of the tendency for consumers to purchase any white paint, not necessarily the best. Dulux decided to remove 'white' from this commodity level and move it towards a positive, conscious decorating decision. It launched Dulux Natural Whites (white paint with a hint of colour) which satisfied the needs of both the white paint users, as well as the 'nervous but would be more adventurous if they could guarantee a safe result' audience. The key to this came from Dulux's recognition that women had become the decision-makers in the decorating market. The advertising and packaging message was aimed at this audience, with impressive results: Dulux's share of the white emulsion market rose from 18 per cent in 1981 to 36 per cent in 1983 (ICI 1985).

The advertising and packaging strategies worked in parallel (see box below right). The advertising stuck rigidly to its long-running but winning formula which included an Old English Sheepdog, which had come to symbolize the quality and reassuring familiarity of the Dulux name. Both the packaging and the TV commercials employed a soft crayon style of imagery that illustrated the product name and benefit. Together they formed the simple sum of the Dulux brand.

The Dulux success was the result of the company understanding the insecurities of its target customer, whilst remaining aware that the brand image represented quality, reassurance and safe adventure. Dulux took the decision to capture a specific market (the hitherto-neglected nervous home improver) simply by asking questions and reacting positively to the response.

Perhaps it is only natural that market leaders, exemplified by Dulux, should move in this way. But what about the second or third-placed brand? Robert Heller (1989) chronicles people and

Proposition
The exciting but risk-free way to transform your home with a new range of gentle whites.

Execution
Soft and reassuring, endorsed by the animated pastels and the emotive branding of the Dulux dog.

Impressions
Confidence – because the paints are white.
Interest – because they are gently coloured.
Reassurance – through the sympathetic naming of Apple White, Rose White, Lily White, Bluebell White, etc.

methods of achievement and states that, according to the law of markets, only the leading two companies and one specialist earn acceptable returns from any market. He goes on to outline a recent product launch which stands the theory on its head.

▎Dry beer

In 1986 Hiroharo Higuchi, a Japanese brewer, took a critical look at his company Asahi, which under its founder had grown in postwar years to provide keen competition to Sapporo and Kirin, the market leaders. After the founder's death, Asahi sales had fallen, along with its market share which dropped to below ten per cent within a year. Higuchi took the view that even an 'also-ran' product had its strengths. On the positive side, he counted the market share, even though it had diminished, and the company's knowledge and expertise accumulated after decades in the brewing industry. On the negative side, Higuchi acknowledged that the company was guilty of complacency and had let itself slip badly. But like many 'also-rans' the company had mainly followed the leaders. Higuchi therefore reasoned that if his competitors were equally guilty of neglect, in time, they too would suffer from a similar inertia, a situation that could possibly be exploited.

Higuchi decided that brewers worldwide believed in their own publicity to the extent that they believed customers would drink whatever they told them to. Higuchi took the opposite view and asked his customers what they really liked and what they could not get. The request was for a beer strong in flavour, high in alcohol, and low in aftertaste.

Higuchi avoided the trap that Coca-Cola had fallen into, and kept his existing beer intact. It was relaunched with a new label and with a new advertising campaign which included a free sampling of one million cans of beer. So positive was the increase in corporate profile for the Asahi product that sales rose by ten per cent. This was, however, only scene-setting to prepare for the March 1987 launch of a new product which Higuchi called a 'dry beer'. The beer's high alcohol level made it marginally less sweet than other beers, but Higuchi emphasized the advantage by naming his product Asahi Super Dry.

As Asahi had only really produced a product to meet the market research brief, the company forecast sales of one million cases, plus a further one million free cans during the remaining nine months of 1987. However, Higuchi had underestimated his own understanding of market psychology. The dual proposition of the advertising, which featured a he-man image, and the clean-tasting product, led to runaway sales of 13.5 million cases. In response, rival companies Kirin and Sapporo both launched similar products. But, instead of cutting into Asahi's market, the consumer's new awareness of these products actually broadened it, and helped to double Asahi's market share in the first six months of 1988.

Asahi Dry Beer: communicating strength and taste through imagery and typography

Much of Asahi's success can be attributed to the positive belief which Higuchi and his team held in their strategy. But it was their willingness to talk to and understand their audience that gave them the initial confidence to adopt such positive positioning. The Asahi proposition was based on the belief in the quality of the product mirroring the conclusive research. The market called for a taste that was specific. Design was used to amplify that message by using bold imagery to communicate strength, and sharp typography to highlight the taste. Advertising played on the same themes through slick, quick-cutting footage of strong-man imagery. The overwhelming impression was that for once, a brewer had created something everyone wanted to try, and the end result, helped on by the free sampling, did not disappoint.

The brand imagery associated with Jordan, Dulux or Asahi involves much more than attractive on-pack graphics. The consumer's perception of how the product should look is more intuitive than they are given credit for. Asked to describe such brands, consumers will list visual elements in order of importance. It may be the colour or the imagery, a detail or a word, but it could also be impressions made up from the sum of the advertising, the pack and the usage together, which form the stronger identity. The successful proposition that has created the advertising and the packaging design to mirror the product's personality, is a valuable property with potential for a long lifespan. If the customer is initially convinced through an expressed benefit of health, versatility, taste or whatever, it establishes an enduring message. A change to the design or advertising strategy must examine the existing image. Even then, the consumer's view must underpin the brief.

5 On-shelf impact

Apart from the contribution made by brand philosophy, advertising history or consumer feeling, design will constitute some fundamental factors of a brand identity. Whether a design 'tweak' or the formulation of a new brief is envisaged, both tasks need to recognize the basic design laws. Even if the ultimate objective is to stimulate the creation of the most innovative design, it should be borne in mind that there is no such thing as an open brief.

As part of the graphic dismantling of a design structure, the starting point will inevitably be the overall impression created by the colour. Understanding the effect that colour has on people will help in making a fundamental packaging decision. Is the product striving for individuality or merely to belong? Almost without exception, one consistent brief requirement is for shelf standout yet, strangely, many products seem still rooted to our accepted idea of programmed colour matching.

▌Colour

As a means of graphic communication, colours are ancient and powerful emotional signals. Originating as symbolic metaphors for the elements, their significance within society extends from religious iconography to everyday safety indicators. Using red and green for traffic lights was not an arbitrary choice. Red has always signified danger, just as green is always connected with naturalness, tranquillity or safety. As young children we react to visual signals before we learn their emotional values, and are most responsive to yellow, white, pink and red. As we mature, our preferential ranking of colour becomes blue, red, green, purple, orange and yellow.

Colour is a precursor to the design brief. Every product sector has its colour rules which, if seeking conformity, it would be foolish to ignore. In practice the art of selecting appropriate colours for packaging produce has developed little from its prepacked heyday in the 1960s. Red equals meat and heat; green signifies 'fresh', from vegetables to toilet cleaner; blue has found its home in the dairy produce sector; and brown continues to project wholesomeness. Putting such generalizations into focus requires no further study than the potato crisp market. The accepted colours of blue for ready-salted, yellow for cheese and onion and turquoise (a synthetic colour for synthetic products) for salt and vinegar have become automatic associations for flavour identification. These simple colour rules guide us through design decisions from evaluating the brief to constructing the proposals.

▮ Standout

If on-shelf visibility (standout) is required for a product then passive colours will be rejected. This is particularly the case with washing detergents. Washing powders have always shouted loudly to grab attention. Not only do they occupy the largest surface area of any supermarket shelf, they also display the brashest colours and the most aggressive typography. This image, however, is at odds with the 1990s' consciousness of the environmentally aware shopper. Detergents are now seen for what they are: harmful pollutants. An ecologically sound product needs to reflect naturalness and care. Since the late 1980s, green has been used to signify these qualities. So the principles on which standout was established are now overturned. As each biological soap powder becomes bolder, it becomes less visible on the shelf in the eyes of the newly aware shopper than its latest sensitively coloured competitors. Accepted colour values and conventional thinking about this product area have been challenged in a way that has considerable implications for the traditional colour values in packaging design.

▮ Typographic style

Product worth is communicated through typography. We recognize, interpret and associate particular typographic styles with quality and function. Typography is a simple science. The letterforms we first formed and read as infants (friendly, rounded lower-case letters) are used to fulfil functions where clarity is paramount. Even if as adults we possess only a scant knowledge of the subject, our familiarity with type patterning is confirmed by daily exposure to newspapers, books, advertising and packaging. Each has prepared us to form and update our opinions. The Gothic script on a newspaper masthead signifies authority and factual reporting. The dynamic hand-drawn typography in comic books communicates action, fantasy and humour. Ornate script on wine labels suggests elegance, sophistication and quality. Traditional Roman serif type signifies maturity, quality and, paradoxically, implies Englishness. Sans serif typefaces communicate with clarity in foggy motorway conditions but seem soulless on hospital waiting room signage. But whether or not a reader can confidently describe the typographic style in which a message is written is of little importance. Moreover, the standard use of typographic styles adds little to the business of creating a unique or memorable brand. It is the dual descriptor of colour and typography that implants the message.

So familiar is the bright red script signature of Kellogg's, that it is relatively unimportant how the company writes the word 'cornflakes' to communicate the sunny values that have been invested in the brand. Similarly, with Oxo it is unimportant whether the typeface that describes the goodness of the ingredients is serif or sans serif, for the red background and round lettering of the Oxo branding is taken to represent meaty taste and versatility. The simpler the supporting text, the greater the emphasis placed on the brand name. This gives the brand its personality and is the focus of the brand's true identity.

The individual 'handwriting' or logotypes adopted by all memorable brands add protection to their name and are part of their unique personality. Our familiarity with the signatures of Kit Kat, Marmite and Persil, to name but three, is so ingrained that a major part of the packaging communication is made by name alone. The real importance is the breadth of individuality that can be established by the adoption of an idiosyncratic signature, rarely achieved by the use of a conventional alphabet. Add to this the dimension of colour and an identity as important as any 'corporate' mark is established.

∎ Naming names

The way a product is received by its audience will be affected by the appropriateness and memorability of its name. As a new product takes shape, the ingredients which make up its identity begin to emerge (see box). How these ingredients interrelate will determine the product's personality and, consequently, what it is trying to communicate and to whom. With such clues, narrowing down choices of colour or typography becomes an easier task.

An early decision governing a product name will be determined by its origins (whether it is to have a corporate, house, range or individual identity), which in turn will depend on the marketing policy to be deployed. With a new product, brand managers have a way of stepping over this issue. Instead of claiming the product as their own, they give it a nickname or a codename. Often this codename will survive throughout the product's development, on some occasions summarizing its qualities through parallel analogy, on others merely seeking to confuse attempts at industrial espionage. Eventually a designer or naming agency will be appointed the task of linguistically anchoring the characteristics that are attached to the new product.

Product formulation
Gives a clue to its potential brand attributes: is it wet or dry, soft or hard, rough or smooth etc?

Physical pack
If transportation or shelf-life is a concern, the physical type of vessel will be important.

Cost of manufacture
The critical equation which positions the product and separates a sale from a snub.

Product function
Signs that will automatically trigger the degree of activity we can expect or must apply to the product. Is it an additive or a preventative? Is it a luxury or a necessity?

The name is a crucial trigger. It communicates a product's personality and it is a name for life. In a few cases the name achieves a more popular abbreviation or nickname; for example, Marks & Spencer is well known as Marks and Sparks, and Woolworths as Woolies. It is essential that the product name, benefit and impression are, at all times, treated as inseparable. Briefing the name independently of the complete visual identity is a mistake.

Where manufacturer and product names complement each other, positive associations are communicated to the consumer

Developing brand names is as strategic and creative as one chooses to make it. Generating dozens of contenders in a brainstorming session or by using a naming agency is only part of the exercise. Manufacturers must register their brand names or guard against name theft as well as ensure that there is no infringement of another company's property. Even the seemingly watertight registered brands of Cartier, Rolex and Louis Vuitton have suffered from flagrant copying in the Far East. Whilst appropriateness to the product is invariably the starting point in most naming exercises, discovering later that a name is registered elsewhere is wasteful of time and money. Many companies have banks of names registered and are waiting either to find a relevant use for them or are using them as a means of blocking competitive advances. Obtaining a licence to use already registered names is possible, but expensive.

Without the support of advertising and promotion, a brand's identity and market share can be eroded by competing products with look-alike names

When Lyons-Tetley launched its Cluster cereal bars in the mid-1980s, it settled on Appleford's as the manufacturer's name in order to deflect any sugary associations with Lyons cakes. The name Appleford's was borrowed from a diabetic food manufacturer, within the Allied Lyons portfolio, which was subsequently renamed Dietade. In this instance, it proved more efficient to rename the original Appleford's company after one of its own brands, thereby saving Lyons-Tetley both time and money in searching for an alternative which could

communicate an equivalent degree of appropriateness.

Name generation can be divided into four categories:

Descriptive

If a product comes from a branded manufacturer, the manufacturer's name will take precedence. It may be asked to perform well across a number of product areas. Homepride, for example, places itself at the risk of imitators and opportunitists with its range of Cook-in-Sauces because there is no copyright protection for product descriptives or statements. So, even if the promise of Homepride products is that they offer better quality than the competition, the company cannot prevent rivals from adopting the same descriptor. The name Homepride linked to Cook-in-Sauce must be sufficiently well established, which means heavy and costly support through advertising and promotion, to withstand the attempts of other products to infringe upon its market share.

Directive

Names which suggest action or activity are a common starting point especially in the detergent and soft drinks sectors – Aim! Direct! Zing! Squirt! and Response!, for example. Although susceptible to copyright infringement, one means of further protecting the name is to register the name in logotype form.

Nonsensical

One of the safest methods of protecting a name is through the use of a nonsense or onomatopoeic word: Bisto (a gravy powder which flavours and thickens all in one go), or Yorkie (surprisingly not a little dog but a chocolate bar), for instance.

Branded

In many cases a product name will rely on an already established and respected brand name. Wall's sausages, pork pies and bacon, for example, are established products from a company that is renowned for its pork-based produce. The idea of Wall's chicken kiev, on the other hand, is less credible due to the lack of poultry expertise associated with the company. Wall's may well be capable of producing excellent alternative products, but it is wise to know your boundaries, for consumers are cautious. They would think twice about Kellogg's beer or Guinness cornflakes. And yet some manufacturers, it seems, can do little wrong and seem to dominate wherever they enter. Heinz, not content with producing rich, savoury products such as their leading tomato ketchup and baked beans, can convince their customers that their baby food is equally palatable albeit aimed at less robust constitutions.

▌Retail choice

Traditionally, it is the branded route which is favoured by retailers who tend to borrow visual identities from one quarter and descriptors from another. As the confidence and credibility of own-brand continue to grow, so too does the number of competitive alternative brands.

Boots, formerly just 'The Chemist', now displays a range of original and independent sub-brands of cosmetics and toiletries such as No.7, No. 19 and the Natural Collection. Similarly, within its own-brand portfolio, Tesco has La Femme, a range of toiletries for women, and Day to Day sanitary towels. Woolworths boasts its Cover Plus paint, and Texas similarly offers paint as a Square Deal.

Retailers have the ability to command on-shelf trial space and they have been frontrunners in blazing the 'green consumer' trail. The launch of Sainsbury's Greencare, a range of washing and cleaning products which, through colour, is disproving the importance of sector language, is set to take a lead in the environmentally friendly branded sector. How long will it be before retailers begin to take the lead in visual branding and manufacturers resort to copycat tactics?

6 Identity parade

Launching a new product, or reassessing an existing one, has increasingly led to the idea that value can always be added through the use of design. Whilst style and quality can be reinforced with a simple adjustment to the basic colour and type, there exists an increasing temptation to enhance the proposition itself with the inclusion of superfluous imagery. Unlike colour and type, illustration is more susceptible to the whims of fashion and subjectivity. Illustration may initially appear to be the vehicle for communicating modernity or a contemporary spirit, but it can quickly date and become a hindrance to the product message. Then the product will require a redesign to keep pace with the 'accepted style'.

The enthusiastic burst of fashionable typography and illustration that heralded the 'design' awareness of the UK's leading grocery multiples seemed to put a refreshing perspective on packaging design. But already it is becoming worn at the edges. Not only has it increased the instances of redesign but it has succeeded in blurring what should be fundamental messages of recognition. Illustration used as a short cut to tap the emotive nerve, has instead resulted in a jumble of imagery that requires unravelling before a positive beneficial identification is confirmed. The initial wave of retail design fireworks sent ripples of panic through the branded manufacturers' ranks. But those who stuck to proven brand identities and remained consistent through both presentation and product have been rewarded by consumer loyalty. Many retailers, on the other hand, have been exposed for their lack of a clear beneficial offer. The consumer is unsure whether they are buying value for money, economy or top quality, because of an over-injection of undisciplined design. Ill-educated, so-called packaging designers, and managers, have employed an over-reliance on fashionable illustration and photographic techniques that have visually confused the consumer. Whilst appearing as an 'image' lift, this misguided approach is no substitute for a clear overall brand identity.

Whilst condemning the flagrant use of 'designer' style, it is important to realize which imagery plays a vital role in a brand's personality. As with most areas of graphic communication, colour and type are the cornerstones of packaging design. They allow configurations of shape and pattern to be constructed that can in themselves become recognizable symbols. The overt graphic approach of Brasso metal polish, with its red rising sun and blue and white sunlight stripes, and the Robertson's Golden Shred story-book label with distinctive three-dimensional type, are just two examples that have been around for a hundred years. Similarly, while Coca-Cola arguably holds the world's attention in typographic messages, it lacks the symbol so recognizable on its rival Pepsi. The red and blue split circle is now a shorthand for comic-style cola recognition, which does not require type to achieve understanding. Similarly, the famous red triangle that is the trademark of Bass beer is so familiar that it is plainly visible amongst the bottles in Manet's 1876 painting of the bar at the Folies-Bergère.

More often than not, the design will include an identifying mark or symbol that has grown to represent the personality of the brand. For example, the implication that Fairy washing-up liquid is 'soft as your face' is well established through the advertising media. Whether the association of 'baby soft' was an idea of the original pack designer we shall probably never know, but a marching baby has been a symbol on the brand for over 50 years, and the relevance this has to the brand's name remains forever questionable. But is it this incongruity that gives many brands an identity, personality and recognition? The following list shows that even through decades of evolution the feature central to the brand's identity remains intact simply through the discipline of retaining the brand's initial message.

Some symbols have obviously grown from the name, such as Houses of Parliament or Swan Vestas, but the majority, like the boot-black kiwi bird, or the 'good morning' cockerel, have been created to endorse the product's benefit.

These characters have little to do with style or with prompting emotion. They are symbols of a brand's individual identity. Each one has a story. For example, Camp coffee (1855) was originally designed for the Gordon Highlanders

Quaker Oats	— *Quaker*
Guinness	— *Harp*
Coleman's	— *Bull*
Robertson's	— *Golly*
Babycham	— *Fawn*
Swan Vesta	— *Swan*
Bassett's	— *Bertie*
Kellogg's Cornflakes	— *Cockerel*
Camp Coffee	— *Highlander and Attendant*
HP Sauce	— *Houses of Parliament*
Typhoo	— *T*
Lyle's Golden Syrup	— *Lion with Bees*
Flora Margarine	— *Sunflower*
Kiwi Shoe Polish	— *Kiwi*

serving in India, and Kiwi shoe polish is an antipodean product. But once their existence is established it is hard to mimic. Who today could convince Lyle's that their golden syrup should carry a dead lion as the feature central to the design, but both he and the accompanying legend ('out of the strong came forth sweetness') are seemingly inseparable from the brand's image.

In the mid-1970s someone, somewhere, instigated the removal of the bull's head from Colman's mustard. Registered as a trademark in 1855, the idea was as radical as removing the Spirit of Ecstasy from the bonnet of a Rolls-Royce car. However, common sense prevailed and in the early 1980s the bull was reinstated, along with the brand's original Victorian typography, perhaps demonstrating that the idea could not be improved upon, even after 125 years. Working with brands that possess such famous marks should be straightforward. The role of the designer is to enhance the benefits of the product and, given the aid of such famous personalities, it requires only the subtleties of typography and layout to reinforce recognition.

What is the relevance of the packaging designer if only such imperceptible changes are proposed? In reality generalization works only at a distance. When looking at any brand in detail, it must become apparent how far a design can move before it loses its identity, or, in a number of cases, before it actually develops a distinct one. The role of the designer is that of visual auditor, assessing the worthwhile messages and stripping out the excess. This approach is not limited to a few really powerful brands whose identities are readily memorable, but includes the less established brands. In the case of younger brands or the brand on the slide, the assessment of visual worth is less easy.

One indication of a brand reaching the end of its natural consumer life is signalled when the brand management team starts to introduce a new design in an attempt to resist the downturn in sales and attract back the customer. Unless the original values of the brand are projected, the possibility of alienating loyal consumers can arise, which can only accelerate the end, thereby reducing any possibility of trading on the brand's existing image for alternative products.

Pure Brilliant White paint – unequivocally Dulux

■ What advertising creates, design reinforces

In certain areas of the market there is little evidence of brand personalities, either because the products have seemed historically dull, or perhaps where the existing brand leaders are so dominant any other might seem a pale imitation. Toilet tissues and vinyl paint, just as recognizable by many as Andrex and Dulux, are prime examples of brands which have personalized seemingly uninteresting product areas. Curiously both utilize dogs for emotive presence. There is, though, a marked contrast between the respective branding policies. While both have successfully used dogs in various advertising themes for the past 20 years, Dulux has acknowledged the real association that its consumers have with the dog and use him as a multi-media endorsement. The Dulux justification for the Old English Sheepdog mascot is that it adds a 'lived-in' quality to the ideal home, and it has subsequently been retained from campaign to campaign for the continuity it gives. Although the original choice of an Old English Sheepdog appears arbitrary and rather more an historical pointer to the popularity of the breed in the 1960s, research has confirmed that it is still well liked by the consumer.

The Andrex puppies satisfy much the same criterion. The playfulness of the puppies allows the opportunity to feature the strength and length of an average toilet roll, and is an overture to imbuing a sensitive area with a human touch. The puppies' dewy-eyed appearance is essentially employed to denote the product's softness – clearly not an association that Dulux wants to encourage for its product.

The strength of both campaigns lies in the easy recall that consumers have when confronted with either the product or the pet. Many people actually refer to Old English Sheepdogs as Dulux dogs, a clear indication of how comfortable they are with the match. It would seem logical, therefore, that having established such a strong rapport with the consumer, the brands should use their pets to the fullest degree.

Dulux, having redefined the white paint market with its hints of white range which kept the company ahead of the flock of product imitators, turned to address the core product of its paint range – Pure Brilliant White. Inevitably when Dulux launched 'Pure Brilliant White' in the mid-1970s the market followed. As a naming exercise it clearly stated an improvement in product quality. However, as a brand name it was vulnerable. Every competitor from Crown to Fads produced its own 'pure' and 'brilliant' version, and by the mid-1980s Dulux was searching for a way of keeping itself further ahead of the competition.

On assessing the working capital that the existing Dulux design possessed it was clear that it retained two important but corporately individual design strengths: first, the reassurance communicated by the dark blue band which carried the Dulux name, and second, the product description which communicated quality in an original and accurate way. The question now was should Dulux rename the product and relaunch it as 'more pure, more brilliant, more white', or was there a way of visually describing these qualities without damaging the simplicity that Dulux stood for? After two years of abortive packaging designs, from room sets to white analogies, and from many design consultancies – all of whom tried to breathe 'originality' onto a paint can – my company, Smith & Milton, was asked to make the can speak for Dulux. We simply put the dog on it. The revised Pure Brilliant White is now unequivocally Dulux, even when the name is not visible. There is no mistaking the manufacturer, nor the quality it offers, and the lifestyle it stands for.

Still on the theme of dogs, the success of petfood market leader Pedigree Chum emanates from its unwavering endorsement of the brand idea. Originally launched as Chum in 1960, it experienced only moderate success against Spiller's Kennomeat. After three years it started to decline, but through exercising the flexibility available to new brands, it changed its brand positioning.

The early years of a brand's establishment are driven by heavy support and one ear on the consumers' reaction. Before the brand image is cast in stone the fine tuning of product formulation, price and positioning can be gently adjusted up or down. Once success strikes, the consumer will come to rely on the product quality and performance every time. The 'new improved' approach is only an option so far.

Having discovered that 50 per cent of users were pedigree dog owners, or at least claimed they were, the Chum brand was relaunched as 'specially formulated for pedigree dogs'. Through the company's discovery that the claim could be substantiated largely by the addition of thiamin (a vitamin sensitive to the pedigree character) and having identified the market for pedigree or 'would-be' pedigree owners, Chum's sales soared. Under the logical symbol of a red rosette, and the appropriately renamed Pedigree Chum, it took over brand leadership by the mid-1960s, a position which it still holds 25 years on with sales in excess of £100 million per annum. The pack design featuring many recognizable breeds, as opposed to the traditional one-dog approach, is simply an extension of the pedigree concept. The endorsement of the annual 'best of show' at Crufts for example, demonstrates how a new direction can be programmed into a consistent advertising campaign. The Pedigree brand name is now so closely tied to the idea of the product that the individual design elements are beyond question: dogs for pedigree confirmation, rosette for prizewinning qualities. For Kennomeat, who had in its control two likeable animated dog characters (Albert and Stanley), nothing of that original strategy nor its market share remains.

This might appear a simplistic view of brand identity but, as is often the case, the more obvious the solution the more applicable it seems. There should be few discernible tricks to packaging design because it is more a specialized form of communication and rarely an artform. Like all communication, once established, recognized and understood, minimal change is required. This, however, is not to advocate an

*Pedigree Chum,
still exploiting the
graphic elements
that gave it brand
leadership in the
1960s*

enforced standstill on brand design, nor to confirm that the 'tweak' is
the best solution. It is my concern that too many brands have destroyed
their original identity by systematic design erosion. Beecham's, for
example, successfully rejuvenated the drink Lucozade without tampering with the original livery or the original brand idea. By repromoting
the product away from hospital bedsides to the drink of medal-winning
athletes, the product has maintained its integrity as an 'energy additive',
but has achieved an edge of activity. Yet Beecham's has continued to
confuse the Ribena identity with new product extensions, adding layer
upon layer of stripe and colour to the original packaging. The

ever-growing range now requires a complete stripping, if the core product message is to regain focus. While the brand has been repositioned in advertising terms from a blackcurrant health drink to a healthier, more modern drink, the packaging has systematically added the 'sugar' the brand was seeking to avoid.

Companies that fail to grasp this simple logic are legion and the opportunities missed extreme. Campbell's, whose tinned soup remains consistently runner-up to Heinz in the UK, is one company which could be accused of undervaluing powerful brand imagery. The simplest design features are often the most memorable yet, while Andy Warhol made the old red and white livery international and legendary, Campbell's persists in ignoring this advantage, preferring instead to show just another bowl of soup.

■ Creating new from old

A recent successful example of brand evolution is Typhoo Tea. Its original dressing was so firmly placed within the design style of its Edwardian origins that in order to keep pace with a changing society and the changing requirements of the tea market (tea bags, foil packs and so on), it has demanded change to signal each advance.

Designed in the early years of this century, the original pack sported a complexity of swirls and scrolls that formed a red letter 'T'. As the century progressed, so the campaigns came and went to endorse Typhoo as the nation's choice. 'Typhoo puts the T in Britain' was one of the most memorable and patriotic campaigns. Throughout this period, the packaging was undergoing change, though still remaining faithful to its Edwardian origins. In 1988, however, Typhoo decided that to stay simple was to remain memorable. Typhoo finally rejected the stereotyped illustration of two cups of tea and two biscuits (an initiative which it had itself instigated, and which has been adopted by many tea brands), and stepped boldly sideways into the limelight. Symptomatic of an action more in keeping with a brand leader (which Typhoo is not) the change was a result of a new and important packaging initiative. By addressing a negative side of the product – tea dust spillage and flavour contamination – the brand was relaunched in foil packs.

As the perception of freshness is critical to the product's appeal, this move has given Typhoo the confidence to take at least the visual stance of a brand leader. Corporately red, with fashioned typography that has evolved not unnaturally from its Edwardian overstatement, it now presents a sleeker, 'quality' look. The advantage of this timeless simplicity is not just a new-found on-shelf visibility, but by shedding the design veneer it reinforces the proposition that Typhoo makes a good, honest cup of tea. We need only the name to jog the advertising memories and a simple symbol of a fresh tea leaf to reassure us that the product remains as fresh as it has always been.

Too few brand identities are maintained with the degree of understanding exercised by Typhoo, Dulux and Pedigree Chum. The history of most brands has gone undocumented, except by those companies which keep a few archive samples that plot the design progression. Where these do exist, the brand invariably remains faithful to its original identity, and in turn to the product proposition. If more brands assessed the visual potential they possess, brand design management could be approached in a far more disciplined and focused way. It is largely irrelevant to research new packaging ideas if the existing pack has not been thoroughly examined. Consumers are far better judges of

In a bold move, Typhoo Tea shed its ornate mantle (right) and achieved quality through simplicity (above)

a brand's personality and the visual elements that go into its make-up than they are given credit for. They are disinterested in brand shares, or sales volumes, and consequently react badly to unsympathetic change. But they do possess the ability to express other key features of a brand's personality often through a more relevant emotional relationship than that of the eager brand manager.

7 Physical packaging

There are very few original products. Most products derive from something occurring naturally or already in the marketplace. It is rare for a new product to make its entry into the marketplace without the need for it to be packaged in some specific way. A product's own make-up dictates its packaging. Success comes when both are appropriate and unique. Take sliced bread. Arguably, the best idea since sliced bread was the packaging itself. In an instant the product was transformed from a generic to a brand communicating freshness and portability. Convenience superseded homebaked freshness. Mother's Pride and its followers precipitated the demise of the small shopkeeper.

In the post-war years there was a conscious effort to shrug off austerity, although rationing continued into the 1950s. The Festival of Britain in 1951 marked a number of refreshing developments, among them the introduction of self-service stores where customers could pick their own goods from the manufacturer's brands on display. The increase in self-service shopping led to products having to show their individuality in order to sell. Here customers found Mother's Pride. The waxed paper-wrapped product was the complete antithesis to Hovis, the only other major bread brand. The branding of Hovis was also unlike that of conventional wrappings, for the Hovis name was baked into the side of the loaf itself. As a brand, Mother's Pride went from strength to strength during the 1960s, and with this success came the need to expand, as much to exploit the brand's advantage as to prolong the product's life. The brand name is now synonymous with the idea of sliced white bread.

A visual and tactile language has grown up around the packaging of produce. Traditionally, the choice for packaged goods was limited and logical: liquids and preserves in glass; dry goods in cardboard; fresh produce in paper; long-life produce in tinplate.

When Mother's Pride adopted a new material it added a distinctive quality to the brand and it demonstrated the practicality and appropriateness of the packaging material chosen. Whilst the choice of packaging materials has increased, the tests of logic and appropriateness still apply:

Visual   – new packaging material to signify innovation
Structural   – sympathetic to the nature of the product
Adaptable   – freshness implied by sealing techniques
Dimensional   – surfaces to provide space for printed messages
Original   – suggestion of a new lifestyle

Whilst it might now seem incongruous to consider sliced bread as a 'lifestyle' product, it was the association of function and convenience that guaranteed its sales. Enough people wanted to try a new and yet very familiar product and this created both the brand and the market. Mother's Pride is just one example of how physical packaging has changed since the 1950s.

Packaging has always been undervalued in its role as a self-promotional vehicle. But the few products that do have a real physical advantage are never slow to advertise their difference. The idea that a

Packaging innovations can contribute both to a product's function and to its appeal

product advantage can be designed into the pack is not a new one, but the instances where this occurs are still rare.

- Jif Lemon is one of the few brands to mimic nature quite unashamedly. It is the closest thing to squeezing real lemon juice, without any wastage.
- Product claims of cleaning 'under the rim' of lavatories have been around for decades. The advantage of cleaning the bowl without having to come into direct contact with it, has accentuated the physical make-up of the Toilet Duck pack. The distinctive shape of the design created the name and added to the product's personality.
- Dulux solid emulsion is a problem-solving product (no paint drips) whose format was tailor-made for the paint roller. There is no escaping the product advantage.
- The toothpaste pump was developed in Sweden. This type of packaging is becoming commonplace for most toothpaste brands. It is a good example of new technology coming to the aid of an outmoded, inefficient system for dispensing paste.

It is difficult to improve upon many traditional methods of packaging, and it is costly for the companies which half-heartedly attempt it. Real technical breakthroughs in packaging are few. The successful new products stand on their own.

In the early 1960s, a Swedish family company, run by Dr Ruben Rausing, developed a new method of packaging liquids. Aware of the worldwide rise of retailers, Rausing looked at how perishable liquids could be sold outside the limited space of refrigeration, without deteriorating. The company developed Tetra Pak, a flexible carton that ascetically packaged liquids after sterilization. The advantage was obvious: without refrigeration, a product's normal life span (even when sealed) of a few days could be extended to several months.

Today, this type of packaging accounts for over 50 per cent of milk sales in Europe, excluding the UK where a doorstep delivery service is still available (Davidson 1987). The major slice of world sales of Tetra Pak is for milk and fruit juice cartons. Only products for human consumption are allowed by the company to be packaged in this way in order to avoid the risk of alienating consumers who may react negatively to seeing detergents or motor oil carried in the same packaging system. The emotional bond that has been carefully constructed to give the consumer confidence in the hygienic standards of Tetra-packaged products remains strong and intact.

Like many product innovations, the success of Tetra Pak was due to the dual benefit of the product. While solving the retailer's need to cut down on expensive refrigerator space, the pack also satisfied the consumer's needs for liquids with longlife properties.

▌Educating the consumer

Unlike changes to pack graphics or colour, physical advances in pack design generally receive little attention from consumers, often because these changes are undetectable. As a result, consumers have been misled into believing that the manufacturer's concern is only with a packet's on-shelf appeal, not its function.

Behind many products' attractive and persuasive exterior lies a direct, though often latent, user benefit. It may be that a product has become easier to use, or that it is better protected against contamination and deterioration, or that it is easier to store, or is tamper-evident, or environmentally friendly, or simply more aesthetically attuned to a modern home interior. Although consumers bought 49 billion Tetra Pak packaged goods in 1988, for example, their convenience, safety and cost benefits are little understood and rarely acknowledged. The same is true of vacuum packaging (where virtually all the atmosphere is removed and the package wall clings skintight to the contours of the product), and modified atmosphere packaging (where an inert atmosphere such as nitrogen is injected into the package to avoid the crushing effects of vacuum). These types of packaging prevent basic foodstuffs like meat, dairy produce, fish, poultry and bakery produce from spoiling; they allow non-foods such as cassette tapes, compact discs and disposable cigarette lighters to be packaged for self-service sale; they ensure that plasters, syringes and other disposable medical items are kept sterile; and they protect high-quality technical products, such as circuit boards, from dust contamination, oxidization, damage and theft.

The lifestyles we now lead, the high standard, self-service retail environments we have come to expect, and the product quality we now demand, owe much to packaging innovations. These innovations have enabled manufacturers to guarantee better and safer products. They have allowed retailers to operate on large and cost-effective scales. And they have provided consumers with opportunities to lead more convenient and healthier lives. But, like Tetra Pak, any such packaging innovation is costly and new developments do not often occur until demand dictates necessity.

In tandem with the brand proposition, benefit will always be the key to success. Totally new packaging systems, such as self-contained ring-pull cans, flip-top cartons and pop'n'press sachets, will always be rare, not least because the cost of embarking upon new product or new packaging development programmes is financially considerable. Perhaps the search for new packaging systems might be better approached from the standpoint of developing and improving existing containers which are readily understood and for which there is already a market, rather than investing in so-called revolutionary ideas, for which a market is currently not available.

USA manufacturers were guarded against innovation in packaging for use in microwaves until ownership reached 40 per cent. This figure has now been reached in the UK and with it has come the development of Freshcap MW, a thermoformed package which remains cool to the touch even when the food inside is boiling. The packs come in a variety of shapes and sizes with the lids incorporating 'easy peel' opening, and, for chilled foods, anti-mist properties. Although easy to open when hot, the lids cannot be peeled back when cold, an important factor which reduces the possibility of product tampering.

But how many microwave users have ever considered that the range, quality and convenience of foods suitable for microwave cooking is determined not by the culinary creativity of food manufacturers, but by the developments of packaging manufacturers? Developments like these make our lives imperceptibly easier because earlier, inferior product designs tend to be shelved the moment new versions are introduced. We quickly forget past examples of packaging. We may moan about the Tetra Pak milk carton being difficult to open without tearing the package or spilling the contents, but we rarely consider the many advantages it has over the glass milk bottle.

With the rise in food sabotage in the 1980s, developments in tamper-evident packaging are taken very seriously by both manufacturers and consumers. Shrink-wrap was an early but inadequate solution because it can be easily steamed off and then re-applied using only a hair-dryer. The 'pop-up' lid devices common to baby food packaging are also highly vulnerable to sabotage. Many designers have been encouraging a move both towards inner packs, which change colour when they come in contact with air, and holograms. Sealed holograms provide a high level of security, are visually attractive, are difficult to reproduce and disintegrate once a package is opened, thereby providing consumers with a warning of tampering. Of course, the significance of this development is reduced considerably if consumers do not understand what the warning means. A survey in the USA showed that 85 per cent of people tested failed to realize that a product had been tampered with when the pack's bottom had been removed (Hattenstone 1989). A programme of consumer education should run hand-in-hand with any developments in tamper-evident packaging.

So far, manufacturers have been slow to introduce such safety initiatives because of the high costs involved. Even with large production runs, holograms, for example, would cost one penny each. Although this is approximately twice the price of the common plastic seal it is infinitely more effective. Perhaps the time has come for manufacturers to recognize that it is more important to protect their products from saboteurs, and so protect their reputation in the eyes of the consumer, than to save pennies.

▌'Going green'

Whilst technologists and packaging experts alike will continue to add longer life and more economy into existing techniques, they could find their efforts superseded by packaging principles once thought redundant, and packaging materials that fail to match their high speed and blemish-free principles. Following a decade (the 1980s) which brought to light many environmental issues which previously were little understood and rarely considered, environmentally sensitive packaging is set to be one of the major growth areas in the 1990s. Already we have seen retailers making great headway and no doubt more manufacturers will begin to realize the selling potential of 'going green'.

The Ark Foundation, set up in 1988, has a range of highly visible cleaning and detergent products which have proved as competitive as the country's leading brands. Consumers have shown their willingness to pay higher prices for a product that offers no apparent improvement in performance. In this instance, quality of product appears to have been replaced by quality of life. Although it is heartening to see an increasing number of retailers embracing environmentally sensitive issues, it is also worrying that many consumers remain confused about the issues and their implications. Many consumers do not know what 'environmentally friendly' means. How green is green? Is the new type of aerosol, for example, environmentally friendly only at the end-user stage, or is it also kinder to the environment during its manufacture? Many points need to be clarified, for in the long run what might be genuine concern on the part of manufacturers, could well be perceived as little more than a push towards self-promotion.

There are now many recyclable and fully biodegradable products on the market. Could a phosphate-free product, in a biodegradable package, which in turn lends itself to a secondary use, and which was produced in a non-harmful manner, be born? In the 1990s, claiming to be green must be all-encompassing: a total, defensible claim.

The relationship between the brand's image and its physical

manifestation can never be discounted: take the easy-to-open shampoo bottle which responds when one is struggling with an eyeful of foam and a scalding shower, or the pleasure in lifting out an individual envelope containing an After Eight mint. Pleasure and performance go together. The success of the Dutch beer Grolsch is due extensively to the consumers' fascination with the stopper and not just to the quality of the product, which could drive other manufacturers to produce more distinctive pack profiles, often at the expense of a real consumer benefit through technological advance.

The instances of combined product and pack satisfaction are too few, and only Tetra Pak stands out as the notable exception, developed over 20 years ago. Like the instances of cosmetic graphics, the real issue of packaging that performs to the benefit of the product has been largely ignored by the manufacturer and left to the packaging innovators who are expected to invent solutions despite the lack of a specific brief. Consequently manufacturers merely buy off the shelf, offer little individuality to consumers, and certainly no unique proposition to their product.

8 Appointing a consultancy

Packaging design is not about pretty pictures and it cannot realistically be tackled by an 'artistic' brand manager and a local printer, however cheap this solution might be. The visual management of a brand is an expert job.

Protecting the brand image can no longer remain the sole charge of the brand manager and the advertising agency. Since the 1980s the balance of brand interest has shifted away from advertising simply establishing a particular proposition and personality for the brand, to rationalizing the total brand imagery to form a stronger base.

Advertising usually demands a high degree of sophistication in the use of image and imagery. But rising costs have prompted advertisers to seek alternative communications media. The marketplace includes large numbers of competing products, many of them own-label. Making a purchase decision based on product performance is increasingly difficult. Packaging can strengthen the brand and create loyalties between the customer and the retailer or manufacturer, a factor that is often overlooked.

Whilst retaining the visual sum of the brand proposition, packaging must reassure through its identification and convince through its imagery. Furthermore, the visual signals and codes that attract the consumer and effect an actual purchase must continue working in the home environment. In essence, a dialogue with the consumer has to be established. Design can add value to a brand as well as encourage emotional responses. This is crucial to building consumer loyalty, and helpful when justifying premium brand prices. Establishing a premium brand requires the ability to interpret the right and not just the most attractive decision. It requires an awareness of the images that are directly relevant to the target market, and a clear understanding of the brand values.

∎ The designer decision

The more professional a design consultancy is, the more it will encourage the marketeer to work in partnership. A good relationship between client and design consultancy is built on mutual confidence and respect. The design consultancy should be entrusted with every piece of relevant brand information, however confidential, in order for it to understand the client's business.

It is common practice for clients to approach two or three design consultancies and expect them to prepare designs based on a common brief, often to a fixed budget. It is a mistake to think this will achieve bountiful results. No small consultancy will be able to respond to the brief with the degree of brand understanding that the client requires, even if they come up with 30 designs. No large consultancy will be able to afford the time to do justice to the brief, even if they considered it worthwhile. Similarly, it is a mistake to believe that the input of three consultancies will result in competitiveness and so a spread of solutions.

The lack of client commitment will inevitably affect the consultancies' input, especially if the work is secured at a low fee. If the client really does need more than half a dozen designs, it probably reflects an unfocused brief, rather than a genuine desire for a number of creative solutions. If the client needs to see a dozen concepts, then a dozen briefs are required in order to achieve a meaningful response.

Track record

The design consultancy should be able to demonstrate its experience in and understanding of branded goods and the branded sector.

Understanding

A good knowledge of the client's markets will probably ensure a fair understanding of its business needs.

It is not essential for the consultancy to have worked within the client's sector. Indeed, if designers do have such experience the client may well not be entirely delighted, for the chances are it was work carried out for a competitor.

People

A client company should establish who they will be working with at the design consultancy, and who will do what. After all, this is to be a team working with and for the client and the chemistry of the individuals involved is fundamental to a successful end result.

Methodology

The design consultancy will have an established way of working. Initially, time is always needed to give the consultancy a clear understanding of the client and the objectives. But extended time scales can result in excessive costs. A consultancy should be flexible in its response to the client's needs.

Implementation

In-house services within a design consultancy range from computer-aided design to mechanical artwork. Any designer can churn out designs whether by hand or machine, but creative thinking takes time, and must be allowed for.

Value for money

A client can always find somebody to take on the brief and do it cheaper. But the client should question whether they are buying the same experience and ability. Often a decision between consultancies is based on price. This may be a mistake if it is a difference of a few thousand pounds on a project worth millions. Good consultancies hire good people, and pay good wages. These people are at the client's disposal, and the client should ask to have their costs explained.

Creativity

A consultancy's credentials presentation of good-looking own-label work does not stack up to a well-presented brand portfolio. What is required in achieving a high level of creativity in the branded sector has long been misunderstood, mostly because a simple presentation conveying a precise message should require minimal 'design'. Look for the projection of brand personality, and make a judgement based on that.

Rapport

Part of the process of choosing a consultancy is talking to the candidates. The more the client reveals, the more the consultancy will understand the project, and vice versa. Meetings that leave the design consultancy unsure of the client's objectives will not expose their skills. If the exercise of awarding a major piece of business was approached in the same way as it is with advertising, the decision making would come from a fairly high level of management. If a brand is important then it deserves time spent appointing the right design consultancy.

The practice

The client company should allow the design consultancy access to all background material and should ensure that senior decision-makers are available for key meetings. Once a project is underway, and despite the autonomy a brand manager may exercise, all parties must be familiar with the design objectives. Likewise, the earlier the design consultancy can be involved in a project, the better its understanding of the work will be, and the less opportunity there is for misinterpretation. No consultancy will be able to make a reasoned comment until a clear understanding of the client's brand is established. Once appointed, the consultancy must be retained throughout the project.

▮ Research

For a brand design to continue working, its appropriateness to the over-all brand strategy must be assessed. In the past, design project briefs were delivered and responded to by instinct, a course of action ill-advised and inappropriate to today's markets. By the use of qualitative research it is possible to discover the positive factors of a brand image, not just a Yes or No. The impressions drawn from the consumer not only establish the visual strengths of a brand, but can be used to confirm the emotive messages which have been built up over time. To begin to understand a brand's equity, this must be a natural course.

As a preliminary step in formulating a new design brief, this assessment of the brand image remains essential. Researching a new design without thoroughly examining the strengths of the existing identity can be costly and time-consuming.

Used effectively, market research can help to focus a design brief by exploring a specific audience's attitudes to a particular market or question. Quantitative research is equally advisable when a major change is proposed or if a project involves a commercial risk.

Research will also reveal the fickle nature of the consumer. Never be surprised to receive a largely negative response to a 'mould-breaking' proposal. Fear of the unfamiliar may draw a negative response. There may be a wide swing of opinion from love to hate. Faced with such polarized views the marketeer invariably exercises caution and starts again. The brave marketeer will back his or her judgement in the belief that the more adventurous respondents are today's opinion makers who will set the standard for tomorrow's followers.

▋ The brief

It is a mistake for clients to allow a degree of 'open-endedness' to the design brief. This is likely to result only in a few fruitless stabs at an answer, with the direction hopefully being established by the resultant design work.

A focused brief should highlight very specific objectives: the business and marketing objectives; the communication strategy; the brand personality profile; production opportunities or guidelines; the creative objectives; research or background material.

If the client fails to give a clear opinion on any of these points, the consultancy will ask for it. The typical response to a brief is for the consultancy to redefine it, in a language tailored to its own needs. This is more a visual brief for the design team and a confirmation to the client that they have grasped the objectives. No matter what the scale of a project the stages of work will follow a trackable pattern (see flowchart).

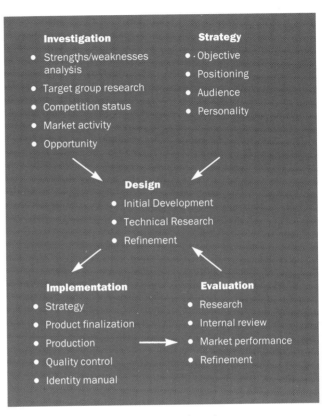

Investigation

- Strengths/weaknesses analysis
- Target group research
- Competition status
- Market activity
- Opportunity

Strategy

- Objective
- Positioning
- Audience
- Personality

Design

- Initial Development
- Technical Research
- Refinement

Implementation

- Strategy
- Product finalization
- Production
- Quality control
- Identity manual

Evaluation

- Research
- Internal review
- Market performance
- Refinement

Investigation

Before attempting any form of communication brief, a thorough understanding of the brand is essential. The relevance of obtaining and understanding consumer response is most important when dealing with a familiar product. One well-used research tool calls for respondents to draw the pack with coloured crayons, from memory. This simple test gives an unerringly accurate account of what visual properties the brand possesses. Any opportunities already spotted or weaknesses suspected, will be further confirmed by this type of study.

Strategy

Having identified an action, the ingredients that form the whole are assembled. The better the understanding of the brand, the more focused the response and the more individual the brand will become.

Design

Design is the visual manifestation of the investigation and the agreed strategy. Following a clear objective the visual vocabulary will be put in place. Lessons learned with type, colour and imagery take the lead. An economy of elements will concentrate the message.

Implementation

At this stage decisions are taken about how the design will work and the time involved to realize it. Whether the design uses only type or photography or illustration, the costs may be high. Only the best resources, materials and techniques should be used. It is a mistake to keep the designer from the printer. In fact, clients should let the designer give them an assessment of the printer's work. Too often, bad preparation by the printer results in a job that is unacceptable either to the designer or the client.

Evaluation

The test of a product's success is its sales. By monitoring profitability and share figures the impact of a new design can be quickly assessed. As the pack is only part of the marketing mix, it is difficult to assess how much it contributes to the product's success. However, if the communications strategy is on course then every other element of the strategy, be it PR or sales promotion, will add to the strength of the brand image.

9 Theory into practice

Putting into practice the outlined stages of a project requires discipline from both parties. Too often the designer is appointed late and merely contributes a graphic 'facelift'. In this instance the designer feels expected to perform, and addresses the problem in a manner that may well be cosmetically pleasing but in the long term fails to enhance the brand character. Too rarely is the advertising agency instructed to sit down with the designer and openly divulge creative and brand strategies. Working with companies which possess the understanding that the packaging and advertising should in fact complement each other is a fulfilling and invariably beneficial practice. The relationship between agency and client is built on openness, a mutual trust, and should reflect a reassurance that each is familiar with the other's object-ives and methods.

▌Natrel Plus

Gillette, the brand most synonymous with shaving, is powerful, American, and markets in over 200 different countries manufacturing products ranging from ballpoint pens to toothbrushes. In the mid-1980s, Gillette held a considerable share of the deodorant market with the brands Rightguard, ZR and Bodymist. Throughout the UK and

Europe the leading brands Rexona, Lynx, 8x4, Narta, Fa and Mum sought to gain market strength through varying propositions such as efficacy, fragrance, and image. Correspondingly the national idiosyncrasies of each market also governed both the advertising approach and the pack graphics. The 'bodily aware' French, appalled at the idea of an anti-perspirant, use only deodorant. The Germans fastidiously avoid any chemical concoctions and favour approaches that major on fresh air and nudism. The fickle British, who primly abhor sweat or odours, either go for maximum protection or apathetically ignore both.

For Gillette, trend research into the diversity of products in the European market was contrary to the narrowing requirements of both sexes in the personal-care product sector. The general move towards quality and fragrance was reducing the differences between products for men and women. The Gillette proposition was to launch a range of deodorant products in Europe, which would appeal to both sexes alike, with fragrance and performance to match. But the main platform for the product was to be 'naturalness'.

Winning the pitch

Four consultancies presented their credentials for the chance to tackle this project. Not insignificantly, it was the advertising agency Saatchi & Saatchi in London who helped Gillette's marketing department to draw up the shortlist. The uniqueness of the proposed 'pan-European' launch meant that none of the competing companies could demonstrate exact knowledge: they had to rely on experience gained at home, or the occasional European sortie. Coming through the first round of meetings, Smith & Milton chose not to represent its track record to the European marketing director, but to present one, well-documented case study, stage by stage. That project, the re-launch of Simplicity sanitary towels for Kimberly Clark, was chosen for the degree of sensitivity required with both the subject matter and its leading brand status. But perhaps more appropriate was the methodology of the working relationship between ourselves and the agency, Ogilvy & Mather. So compatible were the resultant packaging and advertising campaigns, based on the same illustrative style and subject matter (active women), that the consumer was witness to a rare chicken-and-egg brand image. Our willingness to accept the lead role that an agency would take in the Gillette project won the day.

Investigation

At the time nobody had launched an fmcg product simultaneously in several European countries, and Gillette aimed initially at eight: the UK, Ireland, Italy, Norway, Sweden, Finland, Holland, and Spain for January 1988, with Switzerland, France and Germany to follow in March 1989.

Considerable research had been undertaken in each target country to evaluate the present consumers' response to the goods currently on the market, and the reaction towards a more distinct offering. To understand the market ourselves, visits were made to the key markets (coordinated by Saatchi's European representatives) and examples of the competitive products were bought and studied. This further endorsed Gillette's observations of the lacklustre competition. After investigating the bathroom habits of half of Europe, hours of competitive TV commercials were viewed, country by country, which proved to be a perceptive eye-opener to style, sophistication and attitude. Only the Elida Gibbs' brand, Impulse, an all-over body deodorant spray (outside the real target market), showed any continuity in advertising style. Furthermore, Elida Gibbs linked a clever endorsement of the packaging product colour throughout the ads, which added an interesting subconscious dimension. For example, if the featured variety was packaged in blue, then the props and effects in the commercial would take a blue slant. But overall the image of the majority of existing brands we investigated was outdated or indistinct, lacking either subtlety or interest.

Strategy

The objective for the design and advertising was relatively straightforward: to produce a distinctive, quality range of deodorant products, which would appeal to both men and women, and highlight the 24-hour product efficacy and the organic nature of the ingredients (plant extracts).

Whilst assimilating the liberal or parochial attitudes to personal care, all parties were aware of the danger of leading with one cultural stance. Nature was deemed paramount to the brand's personality; the only reservation was that representations of nature could be misinterpreted to suggest a herbal or an overtly floral 'flavour'.

The name

Saatchi had been investigating a name for the brand for several months, aware that English, although important, was not necessarily the only language appropriate to the product. Developing a convincing multilingual name that is appropriate to specific product characteristics is a thankless task. But the name Natura, through both its international sound and its link with nature, was finally chosen. Months after everyone had become comfortable with the look as well as the sound of it, and after some more detailed legal delving, Natura turned out to be registered to several obscure European brands, and was consequently unclaimable. Natura became Natrel.

Design

There were simple visual triggers that would shape the eventual appearance of Natrel, each a significant part of the brand's image.

Natrel: packaging designed for a Europe-wide market

Despite the strong initial leads, a design exercise took place where permutations of colour, type and imagery were put together, as much to eliminate alternatives as to try to alight upon the right one. In the meantime, the advertising agency was also working on its creative brief, one that would simply play to the product's natural strengths.

The design route

It is not unusual to arrive at a design solution by stages. The first round of conceptual work for Natrel produced no single solution, but a central tone of voice that was easily developed into one message, by marrying together the elements that best expressed the desired message.

It was apparent from the first stage that white was to play a dominant role on the pack, as its purity suggested reassurance. A serif typeface, re-drawn to protect the individuality of the name (as yet unchanged), was chosen as an international symbol for quality and sophistication. The scale in which it was reproduced and its vertical positioning made it accessible and striking.

But the key to the brand was its naturalness, and for this we went to the source of the matter and represented a slice of the earth's core, topped and tailed by sun and moon to depict the product's 24-hour efficacy. This stripe of graduated colour visually underlined the simple message.

The stages of refinement – achieving the right balance of elements, adapting the corrected copy descriptors, and so on – were all relatively painless. The decision to respect each country's national pride required the use of different languages for the sub-copy on the front and back of the pack. The name Natrel remained untouched.

The development of three fragranced varieties was an obvious extension and was managed by changing the colour of the title, and adding a coloured dome to the pack lid in order to reinforce the product difference. From the start, the desire to extend a complete image across the whole pack had been expressed, and the distinctive and somewhat 'quirky' appearance of the closure added to the brand's individuality.

Timescale

The initial brief was received in November 1986 and the brand was launched in February 1988. The original timing plan for the design process from investigation, implementation, to the handing over of mechanical artwork, was 21 weeks. Four weeks were spent on the preliminary market study; another four on researching the design, and 25 working days were allowed for the client to discuss and respond to the various stages.

Advertising

In the television commercials the Natrel pack comes to life. From sunrise to moonrise, the natural world and the human race are as one. The human form and natural elements, such as trees and rocks, move together. Empathy between product and purchasers has established the Natrel character. The packaging reinforces the product claim.

Market performance

The European deodorant market is very crowded. The leading brands are heavily backed and command much consumer loyalty. To achieve even 2 per cent of the market share is a considerable task. Within six months of the launch of Natrel, Gillette took a 5 per cent share in Finland and Norway, 4 per cent in the UK, and the target 2.3 per cent in the Netherlands. By the end of 1989 sales were £1.7 million above the targeted figure and a UK trailing study showed a 70 per cent repurchase rate, which is very high. The cumulative effect of product performance, advertising and design has yet to realize its full potential but, as for any new brand, the marketplace offers the best testing ground for compatibility of product and image. The packaging has undergone a slight refinement; there has been a subtle increase in the weight given to the brand name to add strength. Similarly, the advertising has been developed to convey a tougher and more active image, in a bid to secure more male purchasers. Awareness through point of sale or extra product is as important to Natrel as to any brand in its infancy. But the first year is already a success. Natrel is in the right position for the 1990s; the product is effective, reassuring and environmentally safe; its image is green; its packaging and advertising are working even closer together; and its stage is Europe.

Brief

To address the increasing competition between own-label products and Bassett's Liquorice Allsorts.

Background

Geo Bassett is a company which manufactures sweet confectionery. It was founded in 1842. The company originated many famous products such as Liquorice Allsorts, Jelly Babies, Dolly Mixtures, and Sherbet Fountain. Liquorice Allsorts is the leading Bassett's brand with a 53 per cent share of a £27 million market. Bassett's is best known for these Allsorts.

Reason for redesign

- Relentless pressure from own-label lookalike products.
- Low perception of product difference on the part of consumers.
- Insignificant packaging (white carton board) which fails either to support brand status or on-shelf standout. The creators and brand leaders are failing, visually, to claim their top position.
- Lack of perception of the brand's personality at pos.

Brand factors

- Bertie Bassett is a likeable trademark used to represent the product and is used extensively in advertising.
- Liquorice Allsorts are a highly visible product that looks good.
- Bassett's and Liquorice Allsorts have a history to build on.
- The product is not child orientated.
- Liquorice has a distinctive taste, and a large proportion of purchases are accounted for by adults.

Desired personality

- Functional: attractive, highly flavoured assortment of liquorice sweets.
- Emotional: familiar, easy-to-consume product, made to a traditional recipe, by a famous company.

Key design elements

- Attractive illustration of product.
- Product title: The original liquorice allsorts.
- Branding: Bassett's logo and/or development.

Technical restrictions

- Five colours.
- Print process: photogravure.
- Design must be capable of translation on to both flow wrap polypropylene and carton board.
- Varnish available.
- Existing pack dimensions.

Design parameters

- Product is purchased by adults (mass market).
- Overt liquorice connotations, ie black, to be avoided.
- Highlight new flavours and product shape: particularly the addition of a liquorice Bertie.

▮ Bertie Bassett

Occasionally the brief does not match the objective. The Bassett's brief (see left) has been rewritten to avoid any breach of confidentiality, and to accentuate the difference between visual and literal thinking.

On first reading, all the emotional and graphic signals have been identified to produce a more persuasive interpretation of the existing, recessive white pack with an illustrated pile of allsorts and its bland product title. But a brief should be quizzed, dissected and put back together again. The central issue with any brand is the branding. It was the association with the manufacturer's name that Bassett's required. Branding by endorsement when the product is so susceptible to plagiarism is indefensible. Emphasizing 'original' is a sound idea, but meaningless unless qualified. By switching the branding and the product claim, the name negates the necessity of the other. As a title 'Bassett's liquorice allsorts' gives the product a heritage and establishes a route that no retailers could follow.

The idea of the sweet being adult confectionery conjures up a host of guilty eating images, but it was proven that the majority of packets sold were consumed shamelessly. The required illustrative treatment called for tempting the consumer to try the product or, in the case of the addicts, defying them to resist it. Photographing the product in the most mouthwatering way, and popping the product in a striped paper sweet bag, gives the impression of tradition (reminiscent of a product weighed from the jar in the sweetshop and inviting for the hand), and tugs at the consumers' emotions.

The most significant element of the brief was a warning against the use of black. Whether it was thought funereal or stated too heavily an emphasis on liquorice, it was undeniably the most complimentary background to such a photogenic product. The passiveness of the original white was largely a result of the difficulty in achieving a really bright white from carton board or thinly printed polypropylene. Equally, like Natrel, the product needed to jump out at the consumer, especially in the visually active sweetstore environment,

Smith & Milton's design (below) represented a bold and direct challenge to the client's original brief. With a black background and prominence given to the manufacturer's name, the pack achieves greater shelf standout than its earlier counterpart (above)

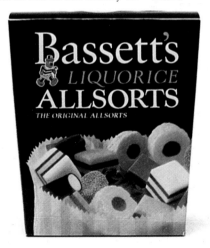

where competition comes in all shapes and sizes. Alternative colours only added contradictory impressions of flavour and were no match for product dimension as it leapt off the black background.

So the brief was effectively turned on its head. The product was renamed 'Bassett's' when the client had asked for 'original', and the pack flaunted black when the client had warned against it. The packaging owes little to the original 1970s pack design, but the message comes through loud and clear and leaves a lasting impression: 'Bassett's have always made the most delicious liquorice allsorts.'

Perhaps my own fondness for the product may have helped me to achieve the desired emotional response, but I prefer to think not. As with all brands the emotional (and beneficial) triggers are already there; it needs understanding on the part of the designer to spring the trap. Since the redesign, Bassett's Liquorice Allsorts' market share has risen to 61.4 per cent.

10 Packaging design in the 1990s

The prospect of the Single European Market has generated much speculation and apprehension about employment and mobility, customs, language barriers, and even whether there is to be a standard plug socket. For the design industry, concern tends to focus on the arrival of the pan-European brand, the brand that will attempt to cut across all social and cultural barriers and appeal to an international market.

For packaging designers, the rules of simplicity which already govern national brands will become even more important when applied to the European and international market. Whilst trade problems with the rest of Europe remain (conflicting business laws, diverse technical standards, and different rates of VAT), few companies have assessed the difficulties or implications of imposing design standards on their neighbours.

The instances of Euro-packs are few. Coca-Cola is the only truly international pack that commands instant recognition wherever it appears. However, the strategy operating behind the Coca-Cola product should not be taken on wholesale by anyone currently contemplating global design. The Coca-Cola image is pure America and does not concern itself with less or more sophisticated image tastes. It is as much a corporate badge as the corporation behind it.

There can be no compromise to the true pan-European or global image. If the sensitivities of one country are less highly developed than the next, then one or the other must adapt according to the drive and personality of the brand, whether new or existing, and one country must take the lead in this process. For example, whilst people in the UK have grown up with Twix and Marathon confectionery bars, the manufacturers, Mars, have dropped these names in favour of their US alternatives, Raider and Snickers. Mars has always adopted a worldwide corporate design stance, so although the Twix and Marathon names have disappeared, the livery will remain comfortably familiar.

What must be avoided is the arrival of Eurobland, the idea that every country will have design reduced to one level, the lowest common denominator. When dealing with countries such as the Netherlands, whose degree of design sophistication in packaging terms is, on the whole, considered to be behind that in the UK, the role of the brand image should be to assist the less sophisticated country, not to restrain the leader. I recently encountered a major brand deciding upon its European creative strategy for the mid–1990s. Two packaging routes were researched. One followed a safe course which featured pleasant but abstract patterns, the other was a watered-down version of proven UK designs. The sedentary route won the day. The result is likely to be that the UK market will receive Eurobland, whereas the Dutch will move forward only slightly, with the rest of Europe similarly following rather than leading. At the same time, the investment in establishing in the UK a strong product personality will be thrown away.

For all manufacturers, the benefits of centralizing production, design and media costs are obvious. Continuity of a brand message is simpler to control if it follows one strategy.

The argument for distinct brand personality is exemplified by the extreme use of recognized sector imagery on the Lever Brothers washing powder, Radion. The pack uses every so-called packaging design cliché in the book, from fluorescent colour to over-emboldened typography, but puts them together in a way that seems to be the definitive washing pack statement. When most product innovation in this area is aimed at tapping environmental awareness, Radion is hard-

selling the more established role of a biological cleaner. So striking is the packaging – polarizing the more fainthearted amongst the design fraternity – that you could not doubt its product claims. Whether British consumers, who have been led gently towards less strident design in the past five years, will embrace the Radion proposition is an interesting question. When the advertising campaign that supported this straight-talking product mirrored the same directness, it seemed initially to be years out of date. But the conventional washing powder format of family dirt, sweat and odour being highlighted and then conquered has proved irrepressible, despite supposedly being hampered by English dubbed onto clearly foreign film. The unmistakable partnership of product, name and pack is linked by the same obvious style which, whether in Tyneside or Toronto, speaks an internationally understood language of washing powders.

In the face of moves towards 'green' packaging by other manufacturers, Radion continues to exploit the sector's traditionally brash graphic language

Radion was launched in the UK in September 1989. It has already claimed 7 per cent of a market worth £620 million. Lever is not shy of admitting that the weight of its £12 million advertising (MEAL) spend has played a significant role. But the impact of the product on shelf and the easy association the consumer has in identifying the product type, make an equally important contribution. What Radion highlights is the ability of design to work largely through symbolic language alone. Radion does not need excessive descriptive copy for endorsement. It is using visual language already clearly understood to do the talking.

▌Sign language

Throughout Europe Gillette is synonymous with shaving products yet, apart from a predominance of simple colours and unified use of the Gillette brand logotype, its packaging for shaving foam maintains little apparent continuity from country to country. No restriction had been applied to the use of sub-branding (such as Foamy in the UK), or to the canister size, which also varies from country to country. If a package has to display several languages, then the text and the graphics need to be simplified.

Similarly, to reduce the number of conflicting existing designs in order to create a common clear message requires an objective approach. Shaving is a hard-edge activity and needs little imagery except for the identification of product type. In this instance Gillette re-grouped the product range and produced seven products, ranging from flavoured foam to a gel. Product distinctiveness was created simply by the use of a symbol which suggested the product benefit: a roundel for example, holding a wavy line for the moisturizing benefit; a jagged symbol for 'heavy beard', the Latin preference. Such simple symbolism was aided only by the brand name, Gillette. The company was confident that the consumer would bring the format and product name together in order to deduce the usage. This introduction of a symbol creates, in effect, a new international language for shaving, aimed at easy identification within a large product range.

The Gillette product has now positioned itself as the international shaving foam with a new identity that demonstrates its successful multilingual communication. One important concession made when addressing the problem of different languages was to produce the can graphics/text in the language of the individual markets. Although adding to the cost, this satisfied the desire for typographic simplicity whilst simultaneously acknowledging the nationalistic pride of each country.

In effect, the confidence displayed by Gillette is synonymous with the power of the 'generic brand'. The decision to leave the consumer to add the word 'shaving' after the Gillette name is an uncompromising move in the company's desire to achieve global success.

There still remains the argument, though, that it is impossible to communicate on such a mass scale when cultural differences and national idiosyncrasies are well defined. There will, of course, always be brands whose appeal is limited to a small and specific cultural audience. However, the case against pan-European brands is weak if the psychology of any brand identity is truly understood. By following a single path of colour, type and symbolism, the brand, if properly promoted, will eventually act like any other understood identity. While differences in climate, food and religion might render certain areas of the international market unsuitable, the opportunities for achieving readily acceptable corporate symbolism are plentiful. One has only to think of the similarity, yet individuality, of national flags, airline liveries and petrol company identities to understand the possibilities for creating European and global branding.

It is unrealistic to expect a design consultancy to achieve international solutions from the insulation of its offices in London. Understanding an international visual identity is not easy and requires knowledge of product area, national preference, cultural distinctions and other intrinsic anomalies. British designers are occasionally appointed by Italian manufacturers, for example, who want to break into the UK market and who require the correct style and tone of voice to make their product credible. However, to employ a UK designer to design for a product for the Italian market requires complete confidence in the designer's international abilities. It is better to apppoint a consultancy on a pan-European basis than on a one-to-one country transfer. Many design companies across Europe will link to provide such a service. Despite the high stakes, the costs of investigation and research are insignificant compared with the potential sales volume of a successful pan-European brand.

Although a pan-European project may be beyond the control of most marketing managers, the qualities and expertise they would seek in a design consultancy differ little from those required when addressing the image of their existing national brands. As the main creative and strategic thrust will, like the marketing control, come from one centre, affiliated design offices located throughout Europe will, like the marketing control, have little to do with the final solution.

■ Environmentally sensitive packaging

During the 1980s the phrases 'environmentally sensitive', 'environmentally friendly', 'ozone layer', 'CFCs' and 'greenhouse effect' penetrated into everyday use. During the 1990s these words and the issues behind them will become even more important and will provide manufacturers with a means of creating benefits for themselves and the environment without necessarily having to search for better products.

There are many opportunities to develop environmentally sensitive products. The implications in terms of on-shelf differentiation and increased sales are vast. The launch and success of the company Ark, amongst others, have already proved the point. Consumers are prepared to substitute performance benefits for environmental well-being without expecting or demanding a reduction in product price. However, the extent to which manufacturers turn the initial disadvantage (large production runs using established methods) into an advantage (the name behind the product) depends on how seriously they view the threat to the environment.

The number of converts to ozone-friendly aerosols, phosphate-free fully-biodegradable detergents and non-animal tested cosmetics is set to rise in the 1990s. But to the average customer there is little tangible evidence of environmental control. Despite the proliferation of 'ozone friendly' symbols, there has yet to be any ruling on their use or on a standardized symbol. In the UK there is little co-operation between government and the manufacturing industries, many of whom are seeking to establish an agreed set of rules. The initiative looks most likely to come from the European government, who in turn could be influenced by the 'Blue Angel' initiative of Germany. Should such guidelines be established, the loopholes that remain in the requirements to display nutritional and ingredient information will need to be closed and the necessary steps needed to comply with manufacturing guidelines could have a greater bearing on the consumer's mind and pocket.

Such initiatives are currently few, particularly in the plastics sector.

Whilst manufacturers continue to mix plastics for ingenuity, it is not feasible, chemically, to recycle the output. Furthermore, of the packaging refuse that is purely plastics based, there is little or no organized system for collection, so again no opportunity to recycle. Consequently there are no buyers for plastics refuse and much of the UK's annual 1.4 million tons of plastics rubbish ends up as landfill. Manufacturers who wish to concentrate on 'greener' issues primarily address their products' ingredients. Greater emphasis must be placed on encouraging recycling bases – on the bottle-bank principle – and consumer education.

▎Physical packaging

A product whose benefit is complemented by its container's physical attributes remains the strongest vehicle for the brand image, yet the least exploited. The cult status achieved by the Dutch beer Grolsch far exceeded the lager connoisseur's expectations. The old-fashioned bottle closure appealed to the image-conscious beer drinker. Similarly, no other confectionary wrapper has come close to the compelling paper envelopes of After Eight chocolate mints, nor will any perfume ever define the visual vocabulary for its sector in the classic manner of Chanel No 5. The 'green' mood must surely open further opportunities to manufacturer and retailer alike. Whilst strong physical presence offers a second point of difference on the shelf, the added value of the guilt-free purchase can tip the balance in its favour. A bottle that is recycled can give a product a packful of added strength: for example a Grolsch bottle is refundable, and the average life of each bottle is some 20 trips. A return to the idea of such traditional packaging methodology, which of course includes the UK milk bottle, should now be uppermost in manufacturers' and packaging producers' minds. Few significant new packaging systems are likely to emerge unless the materials used can be totally justified. 'Green' packaging is set to become commonplace by the end of the decade.

▮ Retail

The 1990s are witnessing a retail slump, although whether the face of the high street will change as significantly as it did during the 1980s is difficult to predict. In terms of product shelf life, many brands and own-brands have already suffered. We are seeing grocery multiples equipped with design managers, image-conscious buyers, and a set of directives about what is and is not right for its personality. The more disciplined identities are emerging, such as the elegant simplicity of Waitrose and the straightforward honesty of Boots. But the consumer is largely left to search for what previously was fundamental high street knowledge: what value does this or that store represent? Design has been applied liberally across the most economic of offerings and it has added a false personality to most of the retail trade. Like the nervous brand using design contrary to its central proposition it could also, ultimately, affect its own future. Who now, by pure visual imagery assessment, can rank by price comparison Asda, Safeway, Gateway, Tesco or Sainsbury's? Retailers must surely revert to attaining a unique identity that reflects the consumer's perception of its true personality. Perhaps only Marks & Spencer, through its relentless search for quality and an uncompromising stance on branded goods, remains distinct. Woolworths has recently begun to exploit the fact that its friendly, low-tech image is held with real affection by the average consumer, and the company has started to capitalize on this by using design to add humour and a smile to shopping.

In terms of in-store design there is currently little to distinguish one store from the next. Standard aisles, standard shelving and standard queues dilute individuality. Perhaps the 1990s will see the re-emergence of corporate rationale and the point of difference will become a total expression for the shopping experience.

But the retailers who followed brands into packaging design and beyond cannot be criticized for their approach to the green issues. It is within their car parks that whole communities must come to turn in their glass and paper for recycling. It is the retailers' products which are most readily adopting 'friendly' materials, as it is they who produce the literature and generate the interest in saving our planet. Their consciousness is starting to go beyond economy. Reconstituted pulp board of the kind used to form egg boxes was phased out for cheaper, more efficient, clear plastics in the 1980s. Now it has been brought back. Glass, so long the *bête noire* of the polished aisle, is once again embraced, and it is retailing advertising which is supplying the average consumer with the conscience to take an active role in responding to its use.

∎ Brand continuity

The issues relating to a strong brand presence will remain central to the growth of packaging throughout the 1990s. As more managers understand the value of a single brand, endorsed by advertising and packaging alike, the temptation to play with the consumer's perceptions will decrease. Brands for the 1990s are in place, honing their image to command on-shelf visibility and shelf life and contemplating the number of possibilities to extend in range or variety. Anyone faced with the sort of issues that have been touched on in this book may find it useful to consider the following thoughts.

Convenience at the cost of brand personality?

If taking stock of brand equity is an essential strategy for the 1990s, it is worth examining a couple of different approaches already operating in British supermarkets.

The first example is of a brand in conflict, torn between its traditional strength, the product characteristic and the desire to add accessibility and further convenience to the existing strong emotional values. Heinz ketchup (it seems unnecessary to add the word 'tomato') was launched at the turn of the century, to form the keystone of the Heinz 57 variety portfolio. The ketchup bottle was so jammed full of natural product that the inconvenience of having to thump the bottle to get the ketchup out became a positive product value. The bottle, whose profile must rank alongside Coca-Cola and Perrier for instant recognition, has always appeared in heavy glass to complement the richness of the ketchup. As the

brand developed, its packaging formats have extended in the conventional way to include different sizes as well as single-portion sachets. But the thick consistency of the product has always remained the identifying brand link. In the mid-1970s an advertising campaign concentrated exclusively on the bottle held upside down and slapped heavily by the palms of various hands. So what prompted the launch of a squeezable bottle in 1988?

Heralded as a great (and rare) technological advance, the development of Lamicon, a multilayered 'squeezable' plastic material, gave Heinz the ideal means of emphasizing the ketchup's distinctive qualities. The Lamicon Heinz bottle was moulded to a flattish shape in order to accentuate its squeezability and given an integral nozzle to control the flow of the ketchup. The convenience of this type of bottle is obvious. But what happened to the brand personality?

If Heinz ketchup becomes the flexible brand, on what platform will the product's worth be supported – on tomato-ness? On better portion control? Or just on squeezability? And if recyclability is to become a major packaging consideration, how can Heinz move in favour of a clearly technological choice against the more environmentally acceptable glass? We have already seen Heinz discard one of its more personable graphic symbols, the 'pickle' that added visual tangibility to the spice of the range. Are we therefore to expect that convenience will rob us further of the features of a favoured brand?

The second product demonstrates consistency and exemplifies the strength of the partnership between a unique pack profile and the product itself. It was the subject of litigation between Reckitt & Colman UK and Borden of the USA. Within its extensive brand portfolio Colman's boasts Jif Lemon, perhaps the closest anybody has come to creating the perfect brand identity. To my mind, packaging design should reflect real life. Its task is to communicate simply and concisely the nature of each product and to take the pain and stress out of shopping. Whilst many packs attempt to establish their personality through conspicuous design, Jif stays as close to nature as is possible synthetically and renders superfluous any further explanation of the product inside.

While the subtle branding of Jif on its 'skin', like Hovis bread, could never stake a claim that it invented lemon juice, Colman's discovered the strengths of associating a natural product with its pack. When Bordern launched a look-alike plastic lemon, the ensuing writ from Colman's was contested on the grounds that there could be no copyright in a natural object (a lemon). The court ruled that the Jif pack was so memorable and individual that Bordern's attempt to trade off the idea was an infringement. Bordern further contested that the coloured cap, flat base and the size of its own pack were sufficiently different from Jif. The court, however, upheld its previous decision, adding further weight to the strength of consumer awareness of the Jif brand.

There are few opportunities to invent brands with such a powerful personality. But getting close to tapping the same emotional nerves is not impossible. So ingrained are our responses to the visual clues that govern brand language, it is possible to construct again and again individual identities that emanate from the same fundamental signals. What is important to grasp is that despite the seemingly complex programme of understanding that is required to manage the visual future of a product, the more that is understood the simpler the programme becomes. Simplicity, in whatever field, is always the most effective route to success.

References

Bernstein, D (1986) *Company Image and Reality.* East Sussex: Holt, Reinehart & Winston.

Booz Allen & Hamilton (1982) *New Product Management for the 1980s.* (Internal publication)

Cecchini, P (1988) *1992: The European Challenge.* Aldershot: Wildwood House.

Davidson, H (1987) *Offensive Marketing.* London: Penguin.

Drucker, P F (1985) *Innovation and Entrepreneurship.* London: Heinemann.

Goldsmith, W and Clutterback, D (1984) *The Winning Streak.* London: Weidenfeld & Nicholson.

Hattenstone, S (1989) Hampering Tampering. *Direction:* October.

Heller, R (1989) *The Decision Makers.* London: Hodder & Stoughton.

Majaro, S (1988) *The Creative Gap.* London: Longman.

Neilson, A C Co Ltd (1989) Britain's Top Grocery Brands. *Checkout*: December.

Olins, W (1989) *The Corporate Identity.* London: Thames & Hudson.

Opie, R (1987) *The Art of the Label.* London: Quarto.

Payne Stracey Partners (1990) *Interbrand Ltd: Brand Accounting Survey.* London.

Peters, T J and Waterman, R H (1982) *In Search of Excellence.* New York: Harper & Row.

Peters, T J and Waterman, R H (1988) *Thriving on Chaos.* London: Macmillan.

Pilditch, J (1987) *Winning Ways.* London: Harper & Row.

Ramsey, W (1982) *The New Product Dilemma.* A C Neilson, Marketing Trend Publications.

Scully, J and Byrne, J A (1987) *Odyssey: Pepsi to Apple.* New York: Harper & Row.

Urban, G L and Hauser, J R (1980) *Design and Marketing of New Products.* New Jersey: Prentice Hall.

Waterman Jr, R H (1988) *The Renewal Factor.* London: Bantam Press.